Quality Out of Control

There is widespread disagreement about what quality in architecture is, and how it can be measured and achieved. Formerly grounded in values of craftsmanship – in the skilled making of things – quality is now associated with the management of administrative or technical processes. Whereas its appreciation was once based on the exercise of individual judgement and taste, increasingly it is founded on supposedly objective systems of evaluation.

Practitioners of design are under pressure to quantify quality, but it is questionable whether it is possible or even desirable to do so. This important and highly topical issue lies at the heart of this book. *Quality Out of Control* considers how, in architectural and cultural practices, 'quality' is defined, evaluated, managed and constructed. It is organised around issues of appreciation, production, belief and measure. With contributions from leading architects and architectural critics, this book is of great interest to architects, clients, architectural academics and students, and anyone interested in what architectural quality is, and how it may be achieved.

Allison Dutoit is an Architect and Lecturer at the Welsh School of Architecture, Cardiff University. She holds a Bachelor of Fine Arts from the University of Michigan and a Masters of Architecture from the University of California, Berkeley. Her research interests in representation and urban design are balanced with practice with Gehl Architects.

Juliet Odgers is an Architect and Lecturer at the Welsh School of Architecture, Cardiff University. As 'Jo Odgers' she is editor (with Adam Sharr and Flora Samuel) of *Primitive: Original Matters in Architecture* (Routledge, 2006) and assistant editor of *arq: Architectural Research Quarterly* (Cambridge University Press).

Adam Sharr is Senior Lecturer at the Wel
University and Principal of Adam Sharr Arcl
Weston) of *arq: Architectural Research Quart*
series editor of *Thinkers for Architects* (Routlec
(MIT Press, 2006; GG, 2008) and *Heidegger for* ... (Routledge, 2007).

Quality Out of Control

Standards for Measuring Architecture

Edited by
Allison Dutoit, Juliet Odgers
and Adam Sharr

Routledge
Taylor & Francis Group
LONDON AND NEW YORK

First published 2010
by Routledge
2 Park Square, Milton Park, Abingdon, Oxon OX14 4RN

Simultaneously published in the USA and Canada
by Routledge
270 Madison Avenue, New York, NY 10016

Routledge is an imprint of the Taylor & Francis Group, an informa business

Typeset in Univers by Florence Production Ltd, Stoodleigh, Devon
Printed and bound in Great Britain by TJ International Ltd, Padstow, Cornwall

British Library Cataloguing in Publication Data
A catalogue record for this book is available from the British Library

Library of Congress Cataloging-in-Publication Data
Quality out of control: standards for measuring architecture/edited by
Allison Dutoit, Juliet Odgers and Adam Sharr.
p. cm.
Includes bibliographical references and index.
1. Architecture. 2. Architectural practice. 3. Building – Quality control.
I. Dutoit, Allison. II. Odgers, Jo. III. Sharr, Adam.
NA2540.Q37 2010
720.1–dc22 2009027299

ISBN10: 0–415–55365–2 (hbk)
ISBN10: 0–415–55366–0 (pbk)
ISBN10: 0–203–86184–4 (ebk)

ISBN13: 978–0–415–55365–0 (hbk)
ISBN13: 978–0–415–55366–7 (pbk)
ISBN13: 978–0–203–86184–1 (ebk)

In memory of James Dutoit

Contents

Illustrations

Figures

Tables

Figure credits

© Hélène Binet: Figures 1.1–1.5
Stephen Kite: Figures 2.1, 2.3 and 2.4
© Ruskin Foundation (Ruskin Library, University of Lancaster): Figure 2.2
Sunand Prasad: Figures 4.1–4.8
Helena Webster: Figures 5.1 and 5.2
Cambridge University Press: Figure 6.1
RIBA Journal: Figures 6.2 and 6.3
Emap Inform, *Architect's Journal*: Figure 6.4
Matthew A. Cohen: Figures 7.1–7.6 and Seminare Archiscoville, Florence, Figure 7.6 inset
Mhairi McVicar: Figures 8.1–8.4
David Leatherbarrow: Figures 9.1–9.4
David Leatherbarrow © FLC/ADAGP, DACS, Paris and London, 2009: Figure 9.5
© Courtesy Takefumi Aida: Figure 10.1
Marc Treib: Figures 10.2–10.12
The Queen's College Archive, Oxford (TQCA): Figure 11.1
Andrew Holmes: Figures 11.2–11.4
Juliet Odgers: Figure 12.1
© All Rights Reserved. The British Library Board. Licence Number: CARUNI16.: Figure 12.2
© Thermae Bath Spa, photographer: Neil Lee: Figure 12.3
© Thermae Bath Spa, photographer: Nick Smith: Figure 12.4
© Thermae Bath Spa, photographer: Matt Cardy: Figure 12.5
Photographs by John Szarkowski supplied courtesy of the estate of the artist and the Pace/MacGill Gallery, New York: Figures 13.1–13.4
Paul Emmons: Figures 14.1 and 14.2
Carol Emmons: Figure 14.3
Nicholas Vella: Figures 15.1–15.3

Contributors

Catherine Belsey has taught at Cambridge and Cardiff, where she chaired the Centre for Critical and Cultural Theory. She is currently Research Professor in English at Swansea University. Her books include: *Critical Practice* (Methuen, 1980, Routledge, 2002); *Desire: Love Stories in Western Culture* (Blackwell, 1994); *Poststructuralism: A Very Short Introduction* (Oxford University Press, 2002); *Culture and the Real* (Routledge, 2005); and *Why Shakespeare?* (Palgrave Macmillan, 2007).

Adam Caruso studied architecture at McGill University in Montreal. He worked for Florian Beigel and Arup Associates prior to establishing his own practice with Peter St John in 1990. Projects by the practice include: New Art Gallery Walsall; Victoria and Albert Museum of Childhood, Brick House, London; and Gagosian Galleries in London and Rome. Nottingham Contemporary and the new Café in Chiswick will open in 2009. The practice has acted as the architects for Tate Britain since 2007.

Hugh Campbell is Professor of Architecture at University College Dublin. His research interests cover a wide spectrum, from Irish architecture and urbanism to the relationship of consciousness and space. His work has appeared in numerous journals and edited collections. Among recent publications are: essays on Hans Scharoun's Philharmonie in *arq*; and James Joyce's Dublin in *Curating Architecture and the City* (edited by Chaplin and Stara, Routledge, 2009). With Nathalie Weadick, he commissioned and curated Ireland's exhibition at the 2008 Venice Biennale entitled 'The Lives of Spaces'.

Matthew A. Cohen is a licensed architect and an Associate Professor of Architecture at Washington State University, Spokane. He holds a Master of Architecture degree from Harvard University, and a Master of Arts in Renaissance Art from Syracuse University. His current research explores new observation-based approaches to the study of architectural history.

Allison Dutoit is an architect and Lecturer at the Welsh School of Architecture, Cardiff University. She holds a Bachelor of Fine Arts degree from the University of Michigan and a Master of Architecture degree from the University of California, Berkeley. Her research interests in representation and urban design are balanced with practice with Gehl Architects.

Paul Emmons is Associate Professor at the Washington-Alexandria Architecture Center of Virginia Tech, where he coordinates the Ph.D. programme in architecture and design. Dr. Emmons, a registered architect, received a Ph.D. from the University of Pennsylvania and a Master of Architecture degree from the University of Minnesota. His research on representation in architectural design practices has been published in anthologies such as: *Body and Building* (MIT Press, 2002); *Aeolian Winds* (Routledge, 2006); and *From Models to Drawings* (Routledge, 2007), and journals including: *arq*; *Journal of Architecture*; *Journal of the Society of Architectural Historians* and *AA Files*.

Stephen Kite is Reader at the Welsh School of Architecture, Cardiff University. His research explores the history and theory of architecture and its connections to visual culture. It has been disseminated through chapters and papers within numerous books and journals, including: *Architectural Research Quarterly*; *Art History*; *Ptah*; *Paideuma*; *The Journal of Architecture*; and *Word and Image*. His most recent publication examines the criticism of *Adrian Stokes: An Architectonic Eye* (Legenda, 2009).

Theodore Landrum studied architecture at Ball State University; practised in New York, Montreal and Ottawa; taught at Norwich University (Vermont); and has been an avid critic at the University of Manitoba. He seeks publishers for two poetry collections (*Tube Traits* and *NY gist*) and is writing another: *Monument to Ephemera*.

David Leatherbarrow is Professor of Architecture and Chairman of the Ph.D. programme at the University of Pennsylvania. He teaches architectural design, as well as the history and theory of architecture, gardens and cities. His recent books include: *Architecture Oriented Otherwise* (Princeton Architectural Press, 2009); *Topographical Stories: Studies in Landscape and Architecture* (University of Pennsylvania Press, 2004); *Surface Architecture* (with Mohsen Mostafavi, MIT Press, 2005); and *Uncommon Ground: Architecture, Technology and Topography* (MIT Press, 2002). Before that were: *The Roots of Architectural Invention: Site, Enclosure and Materials* (Cambridge University Press, 1993); and *On Weathering: The Life of Buildings in Time* (with Mohsen Mostafavi, MIT Press, 1993). His research has focused on selected topics in the history and theory of architecture, most recently on the impact of contemporary technology on architecture.

Mhairi McVicar is Lecturer at the Welsh School of Architecture, Cardiff University. Her Ph.D. research on the theme of precision in architectural practice examines modes of communication between architect, builder and site, as revealed by a series of constructed details.

Juliet Odgers is Lecturer in architecture at the Welsh School of Architecture, Cardiff University. She is currently engaged on a Ph.D. study concerning theories of nature and epistemology in the gardenist work of John Evelyn and is assistant editor of *Architectural Research Quarterly* (*arq*) (Cambridge University Press). As 'Jo Odgers' she is editor of *Primitive: Original Matters in Architecture* (with Adam Sharr and Flora Samuel, Routledge, 2006).

Sunand Prasad is Senior Partner of Penoyre & Prasad and was president of the RIBA (Royal Institute of British Architects) from 2007 to 2009. He founded the Institute's Constructive Change Group and, as vice president, led the Institute's response to climate change. He was a founding Commissioner for the UK Government's Commission for Architecture and the Built Environment and had a leading role in the development of Design Quality Indicators for the Construction Industry Council. Prasad has taught and lectured extensively in the UK and India. His theoretical work includes: research in North Indian urbanism and domestic architecture; writings on Le Corbusier, on architecture and cultural diversity; and the exploration of the value of design. He holds a Ph.D. from the Royal College of Art.

Flora Samuel is Chair in Architecture at the University of Sheffield, where she teaches design and architectural history. She has written extensively on Le Corbusier and is currently writing a book on the Promenade Architecturale for Birkhäuser.

Adam Sharr is Senior Lecturer at the Welsh School of Architecture, Cardiff University, and principal of Adam Sharr Architects. He is: editor (with Richard Weston) of *arq* (Cambridge University Press); series editor of *Thinkers for Architects* (Routledge); and author of *Heidegger's Hut* (MIT Press, 2006; in Spanish with Gustavo Gili, 2008) and *Heidegger for Architects* (Routledge, 2007).

Marc Treib is Professor of Architecture Emeritus at the University of California, Berkeley, practising designer and noted landscape and architectural historian and critic who has published widely on modern and historical subjects in the United States, Japan and Scandinavia. Recent books include: *Settings and Stray Paths: Writings on Landscapes and Gardens* (Taylor & Francis, 2005); *Representing Landscape Architecture* (Taylor & Francis, 2007, editor); *Drawing/Thinking: Confronting and Electronic Age* (Routledge, 2008, editor); *Appropriate: The Houses of Joseph Esherick* (William Stout Publishers, 2008); *Dan Kiley Landscapes: The Poetry of Space* (William Stout Publishers, 2009, co-editor); and *Spatial Recall: Memory in Architecture and Landscape* (Taylor & Francis, 2009, editor).

Igea Troiani is Senior Lecturer at Oxford Brookes University. She has been producing works on the relationship between internal politics and modern architecture since 1999. She has published in *The Journal of Architecture, Fabrications* and *Critical Architecture* (Routledge, 2007), and co-edited *The Politics of Making* (Routledge, 2007).

Helena Webster is an architect and Reader at Oxford Brookes University. Her current research focuses on architectural education, for which she was awarded a National Teaching Fellowship in 2005. She has published extensively on architectural education in academic journals and is currently writing a book, *Bourdieu for Architects.*

Richard Weston is Professor of Architecture at Cardiff University, editor of the refereed journal, *arq*, and director of Richard Weston Studio Ltd. His current work involves translating natural materials into two- and three-dimensional images and materials using digital technologies, as discussed in his most recent book, *Formations: Images from Rocks* (Deukalion, 2007). Previous books include: *Utzon* (Bløndal, 2008), the only monograph written with full access to Jørn Utzon and his archive; *Materials, Form and Architecture* (Yale University Press, 2003); *Alvar Aalto* (Phaidon, 1995), winner of the Sir Banister Fletcher Prize; and *Modernism* (Phaidon, 1996), which received the International Book Award of the American Institute of Architects.

Acknowledgements

The contributions collected here derive from a conference held at the Welsh School of Architecture in Cardiff in July 2007 on the theme of quality. The editors are grateful to all the participants for their contributions, notably those whose work we have not been able to include here. We would like to thank our keynote speakers, in order of their presentations: Richard Weston (who suggested the idea for the conference in the first place), Chantal Brotherton-Ratcliffe, Catherine Belsey, Sunand Prasad, David Leatherbarrow, Adam Caruso, Marc Treib and Beatriz Colomina. Our thanks are also due to the following for their invaluable support in organising the event: Katrina Lewis, Anna Fallows, Phil Jones, Samuel Austin, Phil Henshaw, Mareile Pfannebecker and Edward Wainwright.

Foreword

Richard Weston

It is 2009, and the economic climate feels rather less conducive to pondering the finer points of 'quality' than the distant summer of 2007 when the papers in this book were presented at an international conference in Cardiff. Google the words 'architecture, quality, assurance', however, and up pops something surprising in our financially straitened times – numerous vacancies for 'Quality Assurance Architects'. Not, it turns out, a new line of work for an embattled profession, but a new role in engineering and IT companies where quality has been formally 'assured' for far longer than the messier world of architecture.

I recall first reading about 'quality assurance (QA)' in *The Architects' Journal* in the late 1980s and, uncertain as to its meaning or likely impact on the profession, asked a senior partner in a major civil engineering firm if he knew what it was all about. His words are engraved in my memory:

> Oh, yes, Richard – we've been QA'd for ages. It's essential on large public projects. It's nothing to worry about so long as you understand that it has nothing to do with quality as you or I would understand it. It's all about your filing system.

For me, that *was* something to worry about, filing and all matters administrative never having been my forte. But at least it allayed my fears about QA. The high priests of the arcane world of the British Standard BS5750 didn't have their quality gaze fixed on evaluating the finer points of a cleverly inflected plan, but on the flow of information between the participants in an architectural project. Lousy architecture could be quality assured so long as it was delivered efficiently and accountably.

My scepticism was, I have to admit, grounded in ignorance. In industry, QA had been developed in the United States in the 1930s, and little did I know that the 1990s would be hailed 'The Decade of Quality', with systems of Quality Management, Quality Rating and Improvement Systems and Total Quality Management and their like blossoming on every hand. QA procedures inevitably began to penetrate the groves of academe, where – as in the wider world – they prompted a concern with more familiar understandings of 'quality'. Suddenly, far from doing our jobs well, we had to aspire to 'excellence' in research and teaching, and when mere excellence had become ubiquitous, 'world class' was the order of the day.

It would by cynical to suggest that, amidst the burgeoning quality bureaucracy, there was no concern for – to quote my civil engineering friend – 'quality as you or I would understand it', but it seemed at best a sideshow, a happy by-product of quality management, not its overriding objective. But if we could 'understand' that other quality, surely we could ponder, analyse and write about it – or is what we supposedly intuitively understand really as elusive as Robert Pirsig's oh-so-'sixties musings on 'the art of motorcycle maintenance'? What better topic, then, for a multidisciplinary conference?

The responses, as the contents of this book confirm, were suitably diverse, but interestingly enough included few proposals from the worlds where quality was, arguably, most readily understood – the crafts. We had hoped for some David Pye-like musings on the 'art of workmanship' in various fields but were, perhaps inevitably, disappointed: putting into words knowledge that resides in movements of the body so highly trained as to appear instinctive is not easy. Nor, one might think, is the challenge of quantifying the qualities of a design. But design teachers do that all the time in giving marks or letter-grades to student design projects. Why not extend such 'scoring' to the world of professional design, as the then RIBA president-elect Sunand Prasad argued is essential to promoting better architecture in an increasingly bureaucratised world? Prasad led the writing of 'Design Quality Indicators' (DQIs) (see Chapter 4), and is a persuasive advocate of their usefulness in evaluating architectural quality in an open, accountable way.

But still the doubt persists. Can 'real', 'genuine' quality, the quality 'you and I understand' be reduced to a set of numbers on a standardised score sheet? The problem, as various chapters published here suggest, lies deep in the transition from pre-modern to modern, technocratic societies. In the former, daily life and the making of all manner of things, from utensils to buildings, were governed largely by experience and intuition, whereas, in the contemporary world, we are increasingly forced to put our faith in publicly accountable systems ill-equipped to evaluate 'genuine' architecture, whose complexity lies beyond the control of any individual or system.

There was a further topical thought behind our concern with quality that was touched on by surprisingly few contributors to the conference, namely its potential role in developing a sustainable future. In the wake of the global 'credit crunch', this seems even more urgent. After decades in which economic growth has been based largely on the 'consumption of quantity', might we now be on the threshold of a new era in which the 'cultivation of quality' becomes a defining feature of our economies and societies? In the former, 'design' increasingly became a means of promoting consumption by rendering things rapidly out of fashion or technically obsolete, and quality was assured by devotion to global brands, many of whose owners spent vastly more on advertising and brand management than on design innovation. Cultivating quality 'as you and I understand it' could soon, like 'slow food', mean something rather different and yet strangely familiar: durable, locally distinctive goods and buildings flourishing as a counterpoint to the products of advanced global technologies.

This utopia, if such it be, may be further off than I like to think. But a concern with issues of quality seems certain to come increasingly to the fore as societies dominated by the values of a global economy, of rapid consumption and designed obsolescence come to terms with making more efficient use of resources and rediscover architectural qualities that have been largely occluded by the coarser claims of design.

Introduction

Allison Dutoit, Juliet Odgers and Adam Sharr

In contemporary building wisdom, it is assumed that any project requires the balancing of three key variables: time, cost and quality. The first two – mostly controlled by contractors, project managers and quantity surveyors – are quantified exhaustively: measured in bills, charts and schedules. Described numerically, the appeal of these documents lies in their air of rational authority. They give comfort to professional and institutional clients who perceive reliability in mathematical data. The third variable, quality, is a much more slippery category in the context of contemporary building procurement. It is often constituted as much behind or between the contract documents as within them. It is located as much in the properties of built reality as in any function of process.

This problem of quality is at the centre of architectural culture and is frequently the source of misunderstanding or misapprehension between architects, clients, inhabitants and others in the construction industry. Current industry practices tend to concentrate on means rather than ends, emphasising process. They seek to maximise certainty, minimise risk and ensure the timely and profitable production of space. Many architects, meanwhile, are interested in ends rather than means. They appreciate rich spatial experiences, good proportions, compelling sequences and fine craftsmanship. This central problem of the respective priorities of means and ends views of quality seems to define much about the experience of being an architect in the contemporary world. It also defines many of the values that contemporary society finds expressed in its architecture. In response, this book seeks to explore the problem of quality: how it is determined, by whom and why. This discussion extends beyond the parameters of quality management thinking into the philosophical and theoretical problems that underlie the idea of quality itself.

In Britain at least, architects have ceded much of the management of the building process to others in the construction industry. However, architects tend to remain tacitly in charge of the quality of the built reality of a project. Their drawings and specifications come closest to anticipating build quality. However, their thinking is communicated through production information for a project, set out in drawings and words as a set of standards or performance criteria for particular building elements. As architects understand all too well, such documents are frequently limited, and specifications regularly fail to encapsulate architectural intentions – a problem reviewed by Mhairi McVicar in Chapter 8 on particular qualities of brickwork achieved at Lewerentz's St Peter's Church. Indeed, difficulties with the translation of idea to workmanship to built reality are not just a contemporary problem, as Matthew A. Cohen shows in Chapter 7 in his account of stonework at the Florentine Basilica of San Lorenzo.

For all that process is important, production information and management structures rarely account well for the appreciation of built architecture and for architectural sensibilities more broadly, such as those evoked by Adam Caruso in Chapter 1, or the delight Ruskin perceived in the marble pavements of Verona, explored by Stephen Kite in Chapter 2. Architectural experiences can astonish, console and contain; yet their qualities are largely felt rather than measured. They are primarily subjective rather than objective, intuitive rather than numerical. It is impossible to predict intuitive responses, impossible to write a performance specification for them. It is hard to challenge the benefits of architectural quality, but it is even harder to map its parameters.

There are plenty of architects for whom architectural quality remains ultimately beyond description, for whom any attempt to rationalise it remains meaningless folly. Although this position may have merit, it also has a commercial by-product, serving to promote the notion of architect as a creative genius, and the idea of the signature building, considered by Paul Emmons in Chapter 14. But there are also architects who have hoped to demystify architectural quality. They have sought to bring it within the bounds of the measurable, seeking to bridge the gap in understanding and communication between different actors in the building industry. For them, this is an urgent task. A current champion of this position is Sunand Prasad, who co-authored 'Design Quality Indicators' – which Juliet Odgers and Flora Samuel discuss in Chapter 4 – at Britain's Construction Industry Council. His ideas attempt to engage with a world of industrial building production whose intellectual parameters were established a generation beforehand through the work of Leslie Martin, among others, who attempted to delineate a science of architectural form that is reviewed here by Adam Sharr in Chapter 6. The thinking behind the DQIs also has parallels with the thinking of architectural educators who seek transparent processes, examined by Helena Webster in Chapter 5. These approaches often serve to highlight a mismatch between architects' approaches to architectural quality and culturally dominant ways of thinking.

Many architects would dismiss as pointless attempts at the prescriptive specification of quality embodied in built architecture. Equally, most architects would claim that the practice of architecture is by no means without standards. Those standards, however, are not prescribed but instead kept by the expertise of the professional body, belonging to individual experts and their collective consciousness, rather than specified in lists and numbers. Here, quality belongs with connoisseurship, with judgement and perception. In Chapter 10, Marc Treib expertly locates quality in contemporary architecture in simplicity rather than complexity. In Chapter 9, David Leatherbarrow considers judgements of quality pervasive among architects in terms of qualification, reflecting upon both building configuration and material finishing. Flora Samuel explores the difficult problem of taste in Chapter 3. Judgement is not solely about consensus, though, and in Chapter 11 Igea Troiani considers the vicissitudes of expert debate in connection with James Stirling's Florey Building in Oxford.

Quality, as Leatherbarrow implies, is a philosophical–theoretical matter, as well as a simply architectural one. Cautious of quantification, Catherine Belsey examines in Chapter 15 how psychoanalyst Jacques Lacan was more interested in accounting for the appeal of cultural objects than in assessing their value. Hugh Campbell considers the notion of 'qualia' in studies of consciousness in Chapter 13. Such philosophical thinking must also be seen in historical perspective. Juliet Odgers argues in Chapter 12 that the perception of a need to quantify architectural quality is a symptom of a broadly technocratic contemporary society, exploring the shift from an early modern (seventeenth century) world, in which direct perception, experience and intuition were valued, to a contemporary world that prefers to rely on numbers and systems to guide daily life.

This book contains diverse approaches to quality in architecture from a range of views. It includes contributions from, or about, believers in the quantification of architectural quality and also includes contributions from those who would refute the merits of quantification. Each author locates quality in both product and process, in both means and ends, and questions their relative emphases. As such, the book offers no definitive solution to the problems of architectural quality. We, the editors, do not apologise for this. If you are reading in search of a singular solution to the problem of architectural quality, specifically for a quantitative formula to follow, then this book will demonstrate not only that one does not exist (yet), but also that many architectural commentators believe that project to be difficult or dangerous. It also demonstrates, for better or worse, some of the mechanisms of criticism or connoisseurship that presently stand in its place. Quality control – specifically the idea that quality can be easily controlled – is rarely given much critical attention. Many would like to believe that that architectural quality has readily agreed standards that can be efficiently deployed. We are not so sure. And we have a hunch that this situation may be more an opportunity than a threat. An understanding of the limits of measurement, the limits of production and the limits of connoisseurship can make for richer buildings, more nuanced management processes and more incisive commentary. The discussions of architectural quality contained here describe a diversity of definitions and approaches. Their potential, we hope, is in helping you to negotiate your own position around and beyond them.

Part 1

Appreciation

We know it when we see it. This is the attitude to architectural quality held by many architects, critics and historians. People who take this view believe that they can readily identify a clear organisational diagram resolved into space and structure, spot exquisite details, good materials and rich spatial experiences, and recognise the evidence of a critical position. This ability to identify quality, to sift the well resolved from the poorly conceived, the 'first rank' from the ordinary, is central to the mechanisms of architecture.

There is no single 'right' way to design, no single answer to any architectural problem. Instead, architects are encouraged to hone their individual judgements, to develop the faculties necessary to respond to any given situation with both rigour and creative freedom. On the one hand, this can promote innovation and free thinking, encouraging open minds to exploit opportunities, to take an independent critical position in any given situation. On the other hand, it can also prompt professional introversion, promoting a set of expert values that are shared by few, misaligned with the construction industry and society at large, articulated through a strange, specialised vocabulary.

Architectural appreciation is central to professional expertise, where the demonstration of refined judgement is a badge of distinction. Its parameters are rarely taught explicitly, if they can be taught at all. Instead, they are inculcated through experience, through studio practice and the peer review of criticism. It is remarkable how much consensus there is among professionals about what constitutes 'good' architecture, about what distinguishes a 'good eye'. And non-experts as well as experts are willing and able to identify 'the real thing', to acknowledge a remarkable spatial experience or a finely proportioned edifice. But when one tries to pin down what *it* is, or how *it* works, all too quickly *it* slips away.

This first part of the book aims to explore the appreciation of architectural quality. Adam Caruso, of the practice Caruso St John, exemplifies the expert architect. The practice's buildings – like their Brick House and extension to the Bethnal Green Museum of Childhood in London, or their Walsall Art Gallery – achieve an 'architect's architecture', which has also been widely appreciated by non-architects. Chapter 1 comprises material taken from an interview with Caruso in which he covers a series of themes outlining some of the nuances of architectural judgement, to do with the physical experience of space, craftsmanship, taste and the spirit of innovation. Given that the refinements of design appreciation are not systematically recorded, that architectural culture is passed on to some extent through an oral tradition, it is fitting that these expert values are here recorded by the document of a conversation.

Stephen Kite locates some of the values discussed by Caruso – particularly those concerning the appreciation of fine materials – in a longer trajectory in Chapter 2. He considers the writings of John Ruskin, arguably the consummate appreciator of architecture, and the aesthete Adrian Stokes, examining their feeling for the qualities of stone in Verona.

In response to Caruso and Kite, Flora Samuel reviews the merits of architectural judgement with reference to the writings of French sociologist Pierre Bourdieu in Chapter 3. For Bourdieu, taste and expertise are political weapons employed by individuals and groups to claim authority. They are used to jockey for position, allowing people to demonstrate competitively their superlative education and refinement. Samuel's menu of taste reviews how the expert appreciation of architecture can become a tool for architects to promote their cultural capital and market their services. But, in doing so, she argues, they distance themselves from the people for whom they build.

Chapter 1

Landscapes of experience

Adam Caruso in interview with Adam Sharr and Allison Dutoit

Following his keynote lecture at the Quality conference, Allison Dutoit and Adam Sharr met with Adam Caruso in March 2009 to discuss issues raised by his talk. Referring to his projects and the life of the Caruso St John office, the conversation explored the qualities of experience, quality management processes, innovation and taste. Caruso emphasises the value of architectural judgement and champions its merits.

Adam Caruso (AC): I used to think that the great advantage of working within a more heterogeneous culture is that we no longer have a unitary iconography. The discussion has moved beyond 'I'm right and you're wrong'. But we don't want to go from a situation of singular meaning to no meaning. I think we live in an age where we engage with our world in a slightly different way, which is not to say it doesn't have meaning. This is why I like experiencing contemporary art: you go into the gallery and the experience washes over you. At the Tate Gallery's Turner Prize exhibition, there are response cards that people can fill out. At first, slightly more than 50 per cent of the people say 'it's a load of rubbish'. But then, when the Tate runs short programmes where they actively engage people, most say 'this is really amazing'. They may not like it, but they can see that it's worthwhile and that it has meaning. Experiencing it takes a certain confidence. Architecture is like that. It works in an environmental way that is about experience and spatial relationships, but things also have *meaning*. That's what my lecture at the Quality conference was about. The English landscape, which started as a privileged iconographic discourse, a literary discourse initially, was overtaken by experience.

Adam Sharr (AS): One of the things we were wrestling with in putting together this book from the conference is that, in management vocabulary, which is becoming more and more dominant, the quality of experience is less valued. It has become a process of trying to manage out, or trying to homogenise, things through standard management processes.

AC: I'm hoping that the management thing has peaked. I occasionally get phone calls asking for comments about this topic. Only last week I had a call from a firm of architects who were doing a CABE [the Commission for Architecture and the Built Environment, in the UK] study on what makes a good client. Their idea was that the brief is decisive. They asked me about our Walsall Art Gallery, as it's one of their so-called 'benchmark projects', and they asked if the project had a very good brief. I said that the brief changed through the development of the design, because designing is a very vivid way of illustrating the relationship of things: sometimes one thing has to give, and therefore you have to change the brief. I said that it's the client that's decisive; to have an extraordinary client was the most important thing. The person cut me off! He said, 'we don't have that much time, the deadline is tomorrow, and our thesis is that a brief is the most important thing.'

Allison Dutoit (AD): Give us a quote that works.

AC: Yes!

What I'm talking about is sensitivity. It's about saying *this* is better than *that*, and of course that's the downside of heterogeneity and postmodernism, of relativism. It's necessary to assign value to things. Management culture thinks we don't need to do that, that we can set up processes with criteria and then we can check off the criteria. I was on CABE's review panel for its first four years, and CABE's overall success was all about meeting targets for the year-end report. They could tick the box if they succeeded in meeting a target. One of the boxes was to see a certain number of projects in the press. That was a way of quantifiably assessing the quality of projects, of assessing their success. But, of course, if they're not about relevant or material values, those criteria are vacuous.

There was a moment when management culture was interesting, and there were very dynamic people involved. When IBM was looking at its processes in the late 1950s and 60s, it was probably a pretty exciting time. There was an overarching corporate culture. The architecture was very interesting, but so were the management structures. Saarinen was working with GM and IBM, for example, involved in the built manifestations, the expression of that culture. The management structure was the ethic of the culture and was manifest in the products. Then it became a formula. The change coincides with a loss of confidence in the public sector, where civil servants became politicised, there were challenges to the civil service, and the public sector was in effect privatised. In that context, this form of management by check-list was the perfect thing. Good management consultants, as soon as they began to work for the public sector, lost their way. But still, anything can be interesting when it's intense and being done by intelligent people.

In a country like Switzerland, which is conservative and has a fairly small public sector, everything is done in a way that is argued out and then set. So, when you do a project in Switzerland, there are stages, and each stage is described and the output is clear, and there is no subsequent room for discussion of things that have been previously decided. Maybe the buildings are much more prescribed, but the quality is much higher. However, the same Anglo-Saxon management people are trying to get into Switzerland; they are actually there already.

AS: So, for you, quality is something very singular and intense, and it's about control?

AC: It may be a slight fantasy, what I'm describing in Switzerland, because it's changing, but in this country there's an idea that the competitive tender is efficient. It comes from a belief in the market and maybe also in the innovation of the contractor. The traditional way of working in Germany and Switzerland, which is like a management form, is that you tender with much less information. You agree rates and you get the subcontractors on board. The site is organised by an architect, whose role is to organise the job like a construction manager, and you develop a detailed design with the subcontractors. So you do fewer drawings because the subcontractors do more, and they can really inform the development of the detail work and – because you don't have a main contractor who's lying to you, saying that something doesn't work when actually they don't want to pay for it – the subcontractors are more empowered and they're also proud of the work.

So it is about control, but it's also about efficiency in the process. There has been a belief in markets, which the current state of the world economy has proven are not inherently efficient. There's a point at which markets become baroque and they're not efficient. They're only adding value for certain people. That's what management culture does. It adds value: all these arcane financial products that are making one transaction into four transactions that you can make money on. But it's a fiction. And you make a lot of money until the fiction is exposed.

AS: You were talking earlier about the kind of quality that is to do with visceral experience. But you've also talked about site and empowerment and pride, and management processes and project predictability. Is there actually a correlation between these things?

AC: Management is rather ahistorical and acultural. I am interested in art and architecture that affect you physically, where you do not need to theorise them or read a book about them. The artist Katrina Fritsch dwells on this, asking why everything has to fall into the language trap? Why does it only exist if you can write about it and think about it and connect with it in a literary way? She wants to make work where you go in and it horrifies, or it brings a smile. Then it's not just a one-liner, but you engage with it in different ways. It has a multivalence. But she's also aware of the history of art while she's working. If one is producing, one has to be aware of the history – I don't think you can produce out of thin air.

Your audience should not necessarily need background information. There have been times when I have been on the south bank of the Thames having a sandwich, and someone comes up and points at the Gherkin [Norman Foster's Swiss Re building on the London skyline] and says 'what's that? *That's different*'. Which I think is a condemnation. People ask me my opinion and I say, well, I don't like it. It's just an office building. If it was a museum then maybe it should look different. But why should an office building stick out so much? A sixty-year-old architect should be at the point in his career where he has the graciousness to say, 'this is just an office building'.

AD: There was a judgement there.

AC: It was a commercial judgement. Or it was an ego judgement. I don't think that popularity contests are a way to decide about quality. That's what management culture does. It lets you enumerate the popularity of something, and that's how the government in this country decides what a good building is: by popularity. And a lot of architects play into that. But this undervalues the importance of history and culture.

On the continent, a city plan is a three-dimensional definition of what you can build on each site. So when a developer embarks on building an office building, there is no trying it on, because the volume is given. In Zürich, it is quite sophisticated. Because they arrive at this definition through discussion between the city architecture department, stakeholders and landowners, they make a decision about the volume that will be profitable to develop but that is acceptable and also able to accept those existing buildings which people within the same process have decided should remain. And when an existing building remains, the plan isn't revised after four years!

Sergison Bates did a building in Finsbury Park. There were endless discussions about the brick, and in the end the planners told them that they had to use *this* brick, and the reason they had to use it is, because although *all* of the other buildings were in a different brick, there was one building, a pub, in the particular brick, and planners got it into their head that they had to have *that* kind of brick. But before they started on site, the pub was demolished. The reality in this time of hyper-development that we've just been through is that you don't know what reality is because it's shifting so quickly. It is what everyone wants and it gives you this architecture that is hovering above reality.

AS: In your talk, you spoke about nature and artifice in the English landscape as an idea of a perfected nature. And you talked about the idea of practising in good faith, which I thought was fascinating. It relates to what you were just saying about the Gherkin being different. This idea of practising in good faith, or in the spirit of something, in the spirit of the city and its tradition and its histories, seems relevant. But then doesn't the English landscape also involve a measure of artifice?

AC: Architecture is artifice, isn't it? The English landscape, especially after Kent, is about propriety. When you read Jane Austen, landscape is described as 'pleasant', but pleasant was obviously better than it is now. It was *pleasing*. From a seat in the drawing room there is a very specific view out to the landscape, and the window and the decoration in the drawing room are conducive to having the kinds of chat that you have in the drawing room, and the view outside is also conducive. Then you'll be outside in the landscape where you can go and walk a public walk or take a more intimate walk, maybe even an illicit walk. In this way, there is also a propriety about the relationship of the artifice to the existing landscape. In part, these relationships were determined economically. With Capability Brown and Humphrey Repton, when there was less money around they could move the mountains less, and that imposed a certain discipline. That's why the English landscape is so different from Versailles or the Villa d'Este at Tivoli, the Italian and French baroque traditions. It's about sensibility: social and cultural sensibility, and then the relationship of culture to topography.

But propriety is also dynamic. It's about society. That's what I love about Adolf Loos. His work was so engaged with the cultural *milieu* in Vienna. It was also quite critical and very funny. And he had this amazing feeling about doing a bedroom or a representational domestic room. He didn't do very many public spaces, but he knew the difference between bar and café. And I think he was doing something of that moment in Vienna for *his* kind of people, for his world, this largely Jewish intelligentsia.

I've had conversations with a few people about what's next in thinking about the city, after OMA (Office for Metropolitan Architecture) and late capitalism. The recent crash is hopefully a conclusion, so what's next? And I think it could be about propriety. If you're building in the city, you have to engage the client's ambitions for the project, but also many more people go *past* the building than go *into* the building, which you also have to recognise. It's like designing a park or public space that isn't defined by sports or by retail, which is pretty difficult to do nowadays. The square we did in Kalmar [Stortorget, Kalmar, Sweden] was like that. The project was interesting because there was no change of programme. It was still a square with a cathedral, a town hall, a government office, and none of that changed. We re-*formed* the square, and it became a really pure urban architecture project (Figures 1.1 and 1.2).

AS: You talked about propriety – graciousness and conduciveness – and you describe things you like as involving discipline and rigour. I'm interested in that, and in the relationship with propriety.

AC: They're not contradictory. The square in Kalmar was an invited competition, and we were the only people who talked about the surface of the square. That was what our project was about. The other two schemes had *features*, and then they drew the surface of the square but didn't say anything about it. Our idea for the surface of the square was about what was already there, about removing later additions to return the square to a kind of primitive state using the existing glacial fieldstones.

1.1
Stortorgert, Kalmar, Sweden. Caruso St John Architects with Eva Löfdahl

The really big cut stones there are 200 years old, and they feel like they have been cut by a Cyclops. We wanted to remove the more conventional cut stones, which we could use for other improvements two blocks away. We were getting rid of later layers, which was against all sorts of conservation legislation – in Sweden (and in the UK), the legislation says that each layer is equally significant, and that physical artefacts become a register of social use – but we removed the last 100 or 150 years of the square's material transformation, because that period was largely about the accommodation of the car. There were long discussions with antiquarians. We convinced them, and they were happy in the end. The process was very rigorous. It was rigorous according to a logic that we partially invented. A mythical origin that we were making; an origin the square never had, but that completely engaged its situation today. You might call it a Joseph Beuys environmentalism: a zero energy, zero material, plus/minus that's pure. Rigour involves being clear about what you're doing and then trying to find the formal and technical means so that the ideas can be conveyed as strongly as possible. Building is always tragic because of the amount of compromise. But we're neurotic in trying to control the process. We always try to do a lot with our projects.

You have to be disciplined in order to achieve rigour. I'm a rather neurotic person, Peter is neurotic in his own way, and probably most of our office is quite neurotic. I don't think those ideas – rigour, discipline – suggest that the outcome is reduced or inaccessible. Most of the things that I like have some element of rigour and discipline in them. I think Fischli and Weiss are incredibly rigorous in the film *The Way Things Go*, for example. It's mad, but, boy, did it take a lot of time to do that film. It's rigorous, and the intent is clear.

1.2
Detail, Stortorgert, Kalmar Sweden. Caruso St John Architects with Eva Löfdahl

AD: There's a sense that the rigour of the discipline is embodied in making, not in the discussion of the thing?

AC: I suppose you could write about our buildings in very complicated ways. It's possible. If you really try to engage the situation of the project, it's not contextualism, because the whole constellation of things that make each project can be very distinct, and what remains consistent is our channelling of the neuroses and our rigour.

It's really about interrogation. Architecture is not the technical resolution of details. That's building, and architecture is not that. When people tell you Frank Lloyd Wright's buildings leaked a lot, who cares? Why is that relevant? I'm sure Versailles leaked a lot too. It's a lazy criticism.

AS: The kind of thing that you do in the office – what you called the channelling of neurosis – is a long way from the management idea of quality, of predictability of processes, that we began with. I'm interested in your process.

AC: We're not at all chaotic in the office. We're very organised. We have to be more professional than most architects to get the opportunity to do what we want to do. We are very efficient and productive, but we are not cheap. That said, I think any of our clients would say that we are very good value. They know that Peter and I are totally committed to every single project, which is exhausting. We don't repeat ourselves very much, which is also exhausting.

We generally design our buildings and *then* we start to work with our consultants. We work with very few consultants and we have great relationships with them, but I think we're very difficult to work with because we think that we do the engineering, that we have most of the structural ideas and the servicing ideas. I've had a discussion a few times with a quantity surveyor we work with about why we can't make money. And he said that, well, our jobs are impossible. Sixty per cent are listed buildings, which are really complicated. And we have complicated clients like the V&A and local authorities. Then he said: 'But that's what you want, isn't it?' I guess we're tragically drawn to impossible situations or we really complicate situations.

Our process is concerned with thinking about construction and describing it. The office was once just Peter and I. Before Walsall, I used to do almost all of the working drawings and I loved it. A working drawing to me is incredibly three-dimensional. Through detail and plan and section I can imagine each assembly and the way the different materials are leaning on each other or groaning against each other. I worked for Florian Beigel for a year and I learnt so much. We used to talk about tolerance – what you can do in steel and what you can do in wood. He has a feeling for that, which was a revelation to me. When you think about construction in that way, all materials become equally interesting. Steel and aluminium are so different, and sometimes we use them together. People say you shouldn't, but it's not a problem. You do it because one is bluer, one is warmer, and one is softer. Sometimes, because one is easier, or because the guy who works it is better.

There's a story about [Sigurd] Lewerentz. He was holding some nails, and someone asked him what he was going to do with those nails. He replied, 'I'm not sure, but I'm not just going to nail two pieces of wood together.' Maybe someone made it up, but the idea that you can make materials have the energy to sustain the formal moves is important. It intensifies things.

The Brick House (Figure 1.3) was a pleasure on site because the contractor's foreman was brilliant, and the bosses were good too. We have jobs ten times the size where you tell the contractors what to do and they don't do it. This guy, though, would say 'I have this problem and I've tried these three things; what do you think?' Often he would say 'I think you'll like this one' and usually he was right. Then the process is pleasurable, because you know this person really feels for the work.

AD: Is it about craft, ultimately, what you're describing?

AC: It's not exactly about craft, because it's not about someone chiselling. It's about managing. It's making a decision and having a number of parameters where you can make a decision that may not be what you initially wanted, but you keep closer to the intent than you would if you kept on the original path and had a disaster. We're rigorous in how we get things built, but actually we make formal decisions very intuitively, and sometimes we'll switch our direction because there is unreasonable resistance to something we have proposed, and we find another way.

1.3
Brick House, London. Caruso St John Architects

AS: When you were talking about working in good faith and buildings in the city, it reminded me of Rousseau's social contract, where people give up a little of their freedom to coexist in society. There's also something about a society in building, a good faith among the team of people working on the building and a kind of society that everyone feels a part of.

AC: That sounds a bit pathetic, though. It sounds like 'oh, I like going on site and I like everybody to be happy'. But on a good project people feel proud of it in the end. Construction should be an incredibly dignified job. The values of British society in the past fifteen or twenty years have taken away the status that construction once had. And that's sad. But there are people who clearly still feel proud about what they do. When we have a good relationship on site, it's like a society, but it's not about medieval guilds and handshakes. I'm not *really* interested in merely getting along with the contractor, and I also don't think you design buildings that way.

AS: Like with Lewerentz's nails. Isn't there also a little bit of deliberate perversity now and again?

AC: Not only now and again. Absolutely. If you're going to use something normal, you have to use in a strange way, or else it's just normal.

On the Brick House we did that. It's a more or less impossible site, but we managed. None of the corners are right angles, and we couldn't get special bricks made. Instead we cut them and epoxy-bonded them so they're like specials, and I love that (Figure 1.4). Where we had big beams spanning over big openings,

1.4
Detail, Brick House, London. Caruso St John Architects

the beams are set back, and we glued the brick slips to the beams so you don't see them, or so you just see a brick depth which you know can't possibly span.

That's another important thing. The architecture is never didactic. The Brick House really was made of bricks, and originally had vaults that were brick, but then they obviously went in the wrong direction so you didn't know what the hell they were doing. That was the point. At Walsall, the beams go the 'wrong' way because they're hanging. All of these things are only for the cognoscenti, and for us. But I think the reason we do it is because it gives a small energy to the building.

I love taste. Erudition. That's what I want to have: good taste. Because how could I do something for a client if I don't have this judgement. An elitist is someone who artificially cuts themself off, someone who only inhabits the rarefied cream as they define it. But why wouldn't you want to be able to make correct judgements?

Chapter 2

'Even pavements in Verona are marble'

John Ruskin, Adrian Stokes and the stones of Verona

Stephen Kite

In the founding book of his theory of art – *The Quattro Cento* (1934) – the twentieth-century English aesthete Adrian Stokes (1902–72) records the remarkable fact that 'even pavements in Verona are marble'[1] (see Figure 2.1). Stokes is seen by a number of commentators as heir to John Ruskin (1819–1900), for whom Verona was also the 'beginning of lovely building' owing to this superlative material, 'coloured by iron into a glowing orange, or pale warm red – the peach-blossom marble, of which [the city] is chiefly built'.[2]

Within their succeeding centuries, John Ruskin and Adrian Stokes are among the greatest evocative writers on architecture in the English language; both are masters of *ekphrasis* – the verbal recreation of artwork experience – based on a passionate engagement with the quality of architecture's materiality. They have been compared in Venice – a city admittedly central to both their imaginations – but surprising little in Verona. Yet, in many ways, Verona overrides Venice as Ruskin's 'dearest place in Italy', while Stokes's *The Quattro Cento* (1932) is subtitled *Florence and Verona*, for it is the fiery power of Verona's marble that, Stokes argues, is counter to the lifeless grey *pietra morta* stone of Florence and sustains the tension of early Renaissance, 'Quattro Cento' realisation. Consequently, it can seem strange that Ruskin has given us the three volumes of *The Stones of Venice* (1851, 1853), and Stokes *The Stones of Rimini* (1934), when these cities have no *stones*

2.1 **Verona pavement**

of their own. Venice took its orange-red *broccatello* from Verona (and its white limestone from Istria), while, at Rimini, much of the carving in Leon Battista Alberti's Tempio Malatestiano – celebrated in Stokes's book – is worked out of Verona marble.

In both writers, it is the case that the qualities of the stones of Verona largely inspired the deep response to materiality on which their theories of art and architecture are predicated. At first sight, Ruskin and Stokes seem to have explored two different cities in treading the marble pavements of Verona's historic core. In basic terms, the former sought a Gothic city, the latter an early Renaissance one. But in examining these two sensitive commentators on architecture's quality on their very homing ground, this chapter aims to probe deeper than such stylistic differences might initially suggest.

In Stokes's near erotic feeling for 'the excitements gained from Verona marble',[3] there is a strongly interwoven sense of the presentness of the past: classical Roman archways can have a 'Gothic fierceness',[4] for example, and in turn inspire early Renaissance arabesques. I shall maintain this 'presentness' by interleaving Carlo Scarpa's (1906–78) work through this text as a present commentary on the material qualities expressed by these two art writers.

Stone and water

Stone and water: in one of his poems, written late in life, Stokes recalls a seaside holiday *At Hythe*: 'I hear from bed at Easter / The drawl – drag pebble's hiss / Breakwaters and steep beach'; and remembers his brother: 'Philip my eldest brother still asleep / Unravels stones with hammer, doesn't chatter. / I might content myself with half his dream'.[5] The admired elder brother, Philip Durham Stokes, had been a passionate amateur geologist – he was mortally wounded in World War I, on 4 April 1917, aged 19. Thereafter, Stokes worked doubly hard as if in compensation for the sibling's loss, just as his powerful sense of materiality seems to have been partly driven by a need to atone for the lost brother's love of geology.

Another driving factor was his dystopian vision of Victorian-Edwardian London, evident in his hatred of the immaterial and 'niggard pretension of the [stuccoed] Corinthian pilasters'[6] of the John Nash-inspired Bayswater landscape of his childhood, and the forbidding paternalism of the monuments of Hyde Park – such as the Albert Memorial. Material compensation then for a sense of loss, and competition too – with Ruskin, the great geologist forbear. *Stones of Rimini*, with its eulogy of quattrocento art, is a gauntlet thrown down to the author of *The Stones of Venice* and his Gothic world: Stokes devotes the entire second chapter of this 1934 book to a geological analysis of the nature of limestone. As an aspiring art writer – particularly in the Veneto – he felt oppressed by the looming presence of Ruskin, as a notebook of 1925 reveals: 'I think Ruskin must have been a eunuch although a great man. He lashes me daily, hurls at me *Stones of Venice*',[7] while Ruskin's father, John James Ruskin, 'was fond of saying that his son had been an artist from childhood but a geologist from his infancy'[8] – by the age of twelve, Ruskin had begun a mineralogical dictionary.

Stone and water: Ruskin's diary of 26 May 1869: 'Sleepless night. Up at usual time – saw wet marble and began drawing of St Anastasia façade.'[9] It had been 'wet all [the] day' before, and the rain had brought out the colours of the marble, inspiring Ruskin to study the pilasters on the unfinished west façade of the great Gothic Dominican church. On this damp morning, it was clearly the material qualities of the marbles, their glowing local colour and colour relations that caught Ruskin's eye, rather than the early Renaissance forms of this part of the façade. This study of the base of the façade – 'Marbles at Verona', number 68 in Ruskin's Reference Series (Ashmolean Museum) – illustrates the tonal range of the *Rosso Verona* from palest red to peach blossom.[10]

For his part, in *The Quattro Cento*, Stokes enjoins the traveller who wants to learn of stone colour to

> spend a day in Venice with eyes on the ground [. . .] You will note when you finally sink exhausted at Florian's that the second step of the Procuratie Nuove under whose portico you sit, is made of white Istrian alternating with Verona marble [. . .] or is it reflection of the sunset, so faint are the salmon pink veins, just a glitter, a confused warning, or is it somebody to the right reading the pink *Gazzetino del Sport* which comes out at this hour?'[11]

In this typically bizarre image, Stokes vividly juxtaposes modernity and the colours of weathered stone.

In the Castelvecchio Museum at Verona (1956–73) Carlo Scarpa – a master of such material and temporal juxtaposition – deploys the tonal harmonies of Verona stone as revetment to the sacellum he designed to hold precious Roman and Lombard relics. As on the San Anastasia fragment, drawn by Ruskin, Scarpa plays with chromatic alternations and adjacencies: the most intense chords of colour are the 11-centimetre squares of Rosso Verona, and the chromatic values of the tesserae are intensified by smooth and rough textures.

Eating Verona

To explore further the psychological responses evoked by these vital stones of Verona, there is the revealing letter Ruskin wrote to his father from Verona on 2 June 1852, melding the optical, haptic and oral. In examining Ruskin's close-range looking at architecture, John Unrau surmises how 'as he approaches to within a few feet of a fine [. . .] building, [his] appetite seems to grow':[12]

> There is the strong instinct in me which I cannot analyse – to draw and describe the things I love [. . .] a sort of instinct like that for eating and drinking. I should like to draw all St Mark's – and all this Verona – stone by stone – to eat it all up into my mind – touch by touch'.[13]

Certainly, post-Freudians have been willing to analyse this instinct. In Stokes's *Smooth and Rough* (1951) – a later text strongly influenced by the child-psychoanalyst Melanie Klein – Stokes footnotes this letter of Ruskin's to support 'smooth and rough' textures as 'generic terms of architectural dichotomy [. . .] better able to preserve both the oral and the tactile notions that underlie the visual'. For 'there is a hunger of the eyes, and doubtless [. . .] some permeation of the visual sense, as of touch, by the once all-embracing oral impulse' of infant experience.[14] Verona marble is 'chunky; an eatable', writes Stokes, even equating its dark glow to 'modelling in chocolate where light is reflected luminous'. 'Verona marble is the stone preferred by me for Quattro Cento effect of stone-blossom',[15] for its nature does not allow forms that seem inorganic or 'stuck-on'; they must be coaxed and polished and rounded to finally emerge lustrous from the depths of the stone.[16]

Preceding Stokes's observation that, in Verona marble, 'rounded forms eventuate' is Ruskin's claim that 'sculpture is essentially the production of a pleasant bossiness or roundness of surface', written in the contemplation of the porch of San Zeno at Verona, which offers 'examples of sculpture of every kind, from the flattest incised bas-relief to solid statues, both in marble and bronze'.[17] Scarpa's circular windows on the Banca Popolare di Verona (1974–81) are indeed 'coaxed and polished apertures', with their generous rounded lippings of light Botticini stone, accented at the bottom by a roll, and grooved drainage tongue of red Veronese marble.

Stokes's notebooks confirm the material basis of his theory of art: 'Verona, then, is to be the foil to Florence, the hard red marble to the soft, grey *pietra serena* so easily and so nicely carved in Tuscany';[18] the last point on nicety taunts the modelled facility of Florentine carving. Furthermore, in the published text of *The Quattro Cento*, he writes: 'To return to *pietra morta*. How could this grey stone from Fiesole, this stone without heart, be felt as breaking into life, how emulate the concrete apparitions upon the surfaces of Verona marble?'[19]

San Anastasia – organicism

Both Ruskin and Stokes share organicist strategies of interpretation of material quality, and these naturalisms are worth exploring further both within, and in the context of San Anastasia. We turn again to Ruskin's diaries, to this page of 1851:

> 1851:
> Verona. 30th August – Again, thank God in my beloved piazza St Anastasia: with the little chapel opposite me – its pointed turrets of dark brick now seen against one flaming opening of pale gold in a grey sky of fading storm.[20]

Below this exclamation are some un-annotated lively sketches of an arch and details, which I have identified as studies of the nearby gate of the tower that defends access to the Duomo quarter of the city at the end of the Roman Ponte Pietra – Verona's oldest bridge (Figure 2.2).[21] Ruskin's analyses of this arch tell much about his organicism and his extremely acute scrutiny of detail and difference.

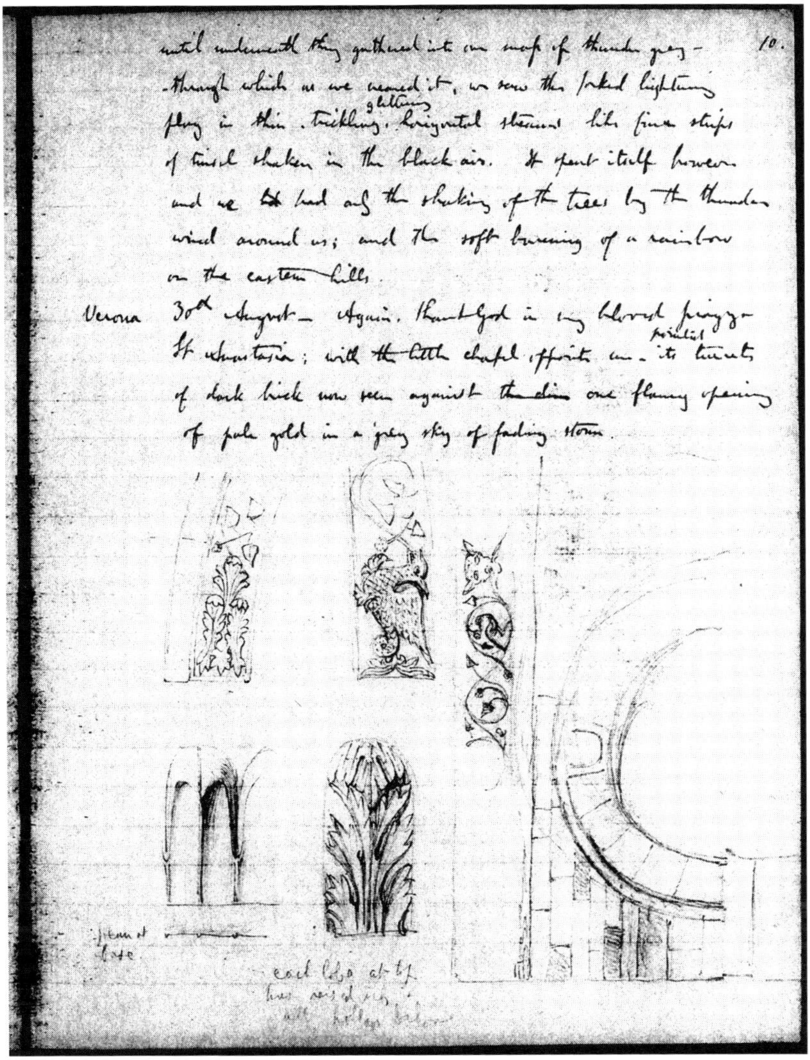
until underneath they gathered into one roof of thunder grey -
-through which as we neared it, we saw the forked lightning
play in thin trickling. glittering horizontal streams like fine strips
of tinsel shaken in the black air. It spent itself however
and we had only the shaking of the trees by the thunder
wind around us; and the soft burning of a rainbow
on the eastern hills.
Verona 30th August – Again, thank God in my beloved piazza
St Anastasia; with the little chapel opposite in its pointed turrets
of dark brick now seen against the dim one flaming opening
of pale gold in a grey sky of fading storm.

2.2
John Ruskin, *Diary 1851–52* (MS 08), Ruskin Foundation (Ruskin Library, University of Lancaster), p. 10

Ruskin had closely studied this gate earlier, in 1849, in one of the very earliest worksheets of the important sequence that was to lead to *The Stones of Venice*. The sternly simple archway is formed of plain, square-cut voussoirs of varied pale and deeper pink stone, borne on imposts carved with a band of vertical, deeply lobed leaves (Figure 2.3). The only other ornament to the arch is the narrow outer border to the extrados, composed of a flush band of flowing tendrils, exquisitely framed by bands of deep rose marble. Ruskin's related diary notes describe how 'the lines of this ornament [of the outer band] and its leaves come gleaming out in silver from the darkened marble: it is impossible that anything could be lovelier or more perfect'.[22] To the casual entrant into the city, who even notices it, it is a graceful, repeating rinceau motif, but Ruskin is alert to the carver's every nuance and variation:

2.3 **Detail of gateway at Ponte Pietra, Verona**

> The *plan* of the moulding is to have a leaf in the large side interstice and berry in small one opposite, but [. . .] a smaller leaf is sometimes as if for experiment substituted, causing a curious and most graceful irregularity.

Then, letting 'l. stand for leaf & b for berry', he carefully tabulates the 'sequence and opposition' in ascending order of the left-hand moulding.[23] The outer edge, for example, begins, as we might expect, in an A/B rhythm of berry/leaf – berry/leaf, but soon goes awry with three repeated leaves and further variants, as Ruskin notes: 'blblblllblbbblbl &c to top'. Meanwhile, at the inner edge, the rhythm of interstices immediately sets off out of step with three repeated leaves before settling down to a berry/leaf alternation: 'lllbllbblblblblb &c to top'.

So, when Ruskin gives a mechanically repetitive, classical Greek running-wave motif as final figure to the first volume of *The Stones of Venice*, he wants to impress on the reader that the true 'mark of nature' is not to be found here – it is inscribed in the living naturalism of work like the endlessly varied and silver-gleaming scrolls of the Verona gateway. He describes the unruly reality of actual waves crashing among the shingle and dismisses the 'great Greek architect' with contempt, who comes 'with measure and rule [. . .] He sets himself orderly to his work, and behold! This is the mark of Nature, and this is the thing into which the great Greek architect improves the sea'.[24]

Ornament and organicism

Entering San Anastasia itself, Ruskin found in Verona's medieval architecture 'churches perfect from porch to apse, with all their carving fresh, their pillars firm, their joints unloosened'.[25] For his part, Stokes 'reserve[s] until the end [of *The Quattro Cento*] the finest of all Quattro Cento incrustation effects', the fifteenth-century reliefs – one of sea-horses, the other of sea-bulls – at the entrance to the Cappella della Jesu adjoining the western side of San Anastasia's south transept.[26] These reliefs fully exhibit that 'intoxication in showing the outer form', that 'joy or completeness in outwardness' by which Stokes defined the 'Quattro Cento'. He draws attention to the undercutting that 'serves Quattro Cento feeling', pointing out 'how the fins which overlap a rim of the candelabra, and the flukes, are cut away razor-sharp'.[27] He concludes with a celebration of this material:

> I would end this book speaking of marble. For once more, nature of the stone is the key to the understanding of the Sant' Anastasia achievement. The reliefs are of the same ivory-grey Verona marble that was procured for the Tempio [Malatestiano, Rimini . . .] and upon which Agostino di Duccio carved his greatest reliefs, those of the planets. This Verona is hard, can be cut to a fine point. It is neither bright nor dull. No other marble is as luminous. Forms sculpted there are distinct yet suffused. The parallel is to objects seen in moonlight and to the curious magnetism of the moon mingling sea and earth.[28]

Ruskin ignores this chapel; his gaze is directed at the Gothic architecture and monuments; Renaissance ones – such as the sixteenth-century Fregoso altar in the first chapel on the right – he dismisses with contempt in his notes as perfect in workmanship but 'wholly without invention, passion or pleasure'.[29] Certainly Stokes accepts that 'Gothic overshadows in Verona'[30] – it is not a Renaissance city. In fact, he breathes deeply the Gothic air of San Anastasia and is intoxicated by it. Indeed, for Stokes, the tension finally achieved by the 'Quattro Cento' is inconceivable without the boost lent by the intense Gothic energy palpable in spaces such as San Anastasia. In the notebook drafts cited earlier, he concludes: 'I need at once to place beyond [Florence] Verona, most Italian, most neglected, of cities, on the other side, from where will awake Pisanello'.[31] And indeed, skied above the Pellegrini Chapel in San Anastasia, is Pisanello's superb fresco of *St George and the Princess*.[32]

But – again reading between both the printed and archive texts of these two great art writers – we can discover one point in this great church where Ruskin and Stokes cease to look past each other; in the presbytery, their gazes finally interlock, with a truly shared enthusiasm, on the fifteenth-century funeral monument to Cortesia Serego (1424–9) (Figure 2.4). It is a revealing point of contact. The mausoleum represents an apex in late Gothic sculptural-decorative naturalism: here, attendants fling back pavilion drapes to disclose the equestrian figure of Cortesia Serego riding upon a shell-niched sarcophagus. Tent and tomb are framed in a great surround of floriated, trilobed arches. Writing in his 'M2' diary of 1850–1, Ruskin seems scarcely able to believe the evidence of his eyes at this near tropical abundance:

> The surrounding crocketing is the most luxuriant I have seen [. . .] It is part house, canopy and all within a huge lozenge or multifoiled figure, composed by a thick trunk of a tree which throws out enormous leaves at intervals inserting the stalks into their centres like some early English monuments or early French.[33]

These rich leaves are 'a middle point between the crocketing of St Marks and the early English, the focus of the latter the feeling of the former'.[34] Here, the reference to the 'feeling' of St Marks is significant, as the crocketing to its west front forms the climax to his sensual recreation of approaching the church – one of the great set-pieces of *The Stones of Venice*:

> Until at last, as if in ecstasy, the crests of the arches break into a marble foam, and toss themselves far into the blue sky in flashes and wreaths of sculpted spray, as if the breakers on the Lido shore had been frost-bound before they fell.[35]

Stokes's 'Quattro Cento' is characterised by an equal erotic tension; he is also amazed by the 'rollicking foliage', which he relates to the Tempio Malatestiano at

2.4 **Equestrian funerary monument of Cortesia Sarego (1424–9) (attrib. Pietro Lamberti or Nanno di Bartolo?)**

Rimini. He further relates this Serego monument to the similar Nicolò Brenzoni (1424–6) monument by Nanni di Bartolo in the nearby church of San Fermo, where – in a comparable instantaneity – putti throw back the curtains of the baldachino to show the moment of the Resurrection. In these monuments that demonstrate the 'passing from Gothic to full Renaissance: a Quattro Cento intensity [is] opened out for us', claims Stokes.[36]

So intense is Stokes's sense of such expression as an efflorescence of material that he goes so far as to reject any ascription of the word 'ornament' to the 'Quattro Cento': 'The quattro cento did not *ornament*', his notebooks insist.[37] In his 1929 essay on 'The Sculptor Agostino di Duccio', he records

> dolphins [. . .] swimming out of the brick above the windows of the palazzo Lettimi at Rimini, a work of the school of Francesco di Giorgio [. . .] The dolphins are swimming out of the brick. Anybody should see that they are not ornament.[38]

While in Verona, he discovers other such fantastic dolphins coiled on the Albergo Accademia, just off the Via Mazzini, whose details exhibit 'a tense communion with the plane which shows them off. They never give an effect of having been put there, just like that. On the contrary they are integral with the plane.'[39] When Stokes writes that the quattrocento 'did not *ornament*', he clearly targets Ruskin's claim that '*ornamentation is the principal part of architecture*'.[40]

Other buildings in Verona, such as the Loggia del Consiglio (1476–92), extend into architecture the sculptural quattrocento attainments of San Anastasia in achieving for Stokes 'an expression of density in the stone, not rhythmic certainly, but firm, solid, tense, an expression on huge and abstract scale of the love the sculptors showed in carving'.[41] Here, Stokes maintains that, 'only if wall-space *means* so much, can this Quattro Cento effect be emulated. A hundred elements of design, of materials and of their treatment, go to make wall-space so pregnant'.

Surrounded, as we are, by expanses of unemblematic cladding and curtain walling, wall space rarely seems so 'pregnant'. Yet the façade of Scarpa's Banco Popolare di Verona – just a short walk from the Loggia del Consiglio – attains an equal tension: its wall-plane stretches tensely between charged apertures and establishes a rich resonance with the fabric of the historic core.

Acknowledgements

I am grateful to Ann Stokes and Adrian Stokes's literary executors, Ian Angus and Telfer Stokes, for kind permission to quote from Adrian Stokes material held in the Tate Gallery Archive. Stephen Wildman, Rebecca Patterson and Diane Tyler offered valuable assistance in archive research at the Ruskin Library. I am indebted to The Paul Mellon Centre for Studies in British Art for grant support towards archive research and fieldwork in Verona and Venice.

Notes

1 Lawrence Gowing (ed.), *The Critical Writings of Adrian Stokes, Vol. 1 1930–1937; Vol. 2 1937–1958; Vol. 3 1955–1967*, London: Thames & Hudson, 1978. Vol. 1, *The Quattro Cento*, p. 179. Subsequent references in the style: *CW*, 1, *The Quattro Cento*, p. 179.

2 E.T. Cook and Alexander Wedderburn (eds), *Works of John Ruskin, Vol. 19*, London: George Allen, 1904–13, p. 342. Subsequent references in the style: *Works*, 19, p. 342.

3 *CW*, 1, *The Quattro Cento*, p. 78.

4 *CW*, 1, *The Quattro Cento*, p. 179.

5 Adrian Stokes, *With All the Views. The Collected Poems of Adrian Stokes*, Manchester: Carcanet New Press, 1981, p. 30.

6 Adrian Stokes, *Sunrise in the West*, New York and London: Harper & Brothers, 1927, p. ix. For further discussion of this Hyde Park landscape, see also Stephen Kite, 'The Urban Landscape of Hyde Park: Adrian Stokes, Conrad and the *topos* of Negation', *Art History*, June 2000, vol. 23, no. 2, pp. 205–32.

7 Tate Gallery Archive (TGA), 8816, notebook 7.

8 Quoted in Tim Hilton, *John Ruskin. The Early Years 1819–1859*, New Haven, CT and London: Yale University Press, 1985, p. 17.

9 E.T. Cook, *The Life of John Ruskin, Vol. 2 1860–1900*, London: George Allen, 1912, p. 160.

10 See Nicholas Penny, *Ruskin's Drawings*, Oxford: Phaidon, 1989, pp. 70–1.

11 *CW*, 1, *The Quattro Cento*, p. 38.

12 John Unrau, *Looking at Architecture with Ruskin*, London: Thames & Hudson, 1978, p. 119.

13 John Lewis Bradley, *Ruskin's Letters from Venice 1851–1852*, New Haven, CT: Yale University Press, 1955, p. 293.

14 *CW*, 2, *Smooth and Rough*, p. 243.

15 *CW*, 1, *The Quattro Cento*, pp. 78–9.

16 For Stokes, 'Quattro Cento art is conceived to be the kernel of fifteenth century Italian art in its aspect of a dramatic release of the human spirit. There is, then, an intoxication in showing the outer form, in stabilizing an inner content as concrete shapes, a joy or completeness in outwardness that is not met with elsewhere.' *CW*, 2, *Venice*, p. 308.

17 *Works*, 20, p. 214.

18 TGA, 8816, notebook 4, in a notebook draft dated 1927, headed, *Quattro Cento Sculpture and Architecture. Florence and Verona.*

19 *CW*, 1, *The Quattro Cento*, p. 103.

20 John Ruskin, *Diary 1851–52* (MS 08), University of Lancaster: Ruskin Foundation, Ruskin Library, p. 10.

21 To my knowledge, these sketches have only been identified to date as 'likely to be Verona subjects'. See Stephen Wildman, *'My Dearest Place in Italy': Ruskin in Verona*, University of Lancaster: Ruskin Library, 2005, p. 31 (exhibition catalogue).

22 John Ruskin, *Diary 1849–50, Notebook M*, University of Lancaster: Ruskin Foundation, Ruskin Library, p. 28.

23 John Ruskin, *Diary 1849–50, Notebook M*, University of Lancaster: Ruskin Foundation, Ruskin Library, p. 29.

24 *Works*, 9, p. 409.

25 *Works*, 16, p. 66.

26 Noting the quality of these 'exquisite reliefs', Stephen Bann comments that 'one of the pleasures of reading Stokes is, then, that it leads you to places where you feel empowered to re-enact the discoveries that he once made'. Stephen Bann, 'Foreword' in Adrian Stokes, *The Quattro Cento and Stones of Rimini*, University Park, PA: The Pennsylvania State University Press, 2002, ix–xvi (p. x). Ruskin has a similar impact, best recorded by Proust.

27 *CW*, 1, *The Quattro Cento*, p. 179.

28 *CW*, 1, *The Quattro Cento*, p. 179.

29 John Ruskin, *Diary 1869–74* (MS 16), University of Lancaster: Ruskin Foundation, Ruskin Library, p. 5.

30 *CW*, 1, *The Quattro Cento*, p. 179.

31 TGA, 8816, notebook 4.

32 An essay on Pisanello, written in 1928, was the first of Stokes's mature art writings. Adrian Stokes, 'Pisanello: First of Four Essays on the Tempio Malatestiano at Rimini', edited and introduced by Richard Read, in E.S. Shaffer (ed.), *Comparative Criticism: Walter Pater and the Culture of the* fin-de-siècle, Cambridge: Cambridge University Press, 1995, pp. 161–207 (pp. 183–4). See also Stephen Kite, *Adrian Stokes. An Architectonic Eye*, London: Legenda, 2009, Chapter 4, pp. 80ff.

33 John Ruskin, *Architectural Notebook, Italy, 1850–51* (M2), The Beinecke Rare Book and Manuscript Library, Yale University Library, p. 123.

34 John Ruskin, *Architectural Notebook, Italy, 1850–51* (M2), The Beinecke Rare Book and Manuscript Library, Yale University Library, p. 123.

35 *Works*, 10, p. 83.

36 *CW*, 1, *The Quattro Cento*, p. 178.

37 TGA, 8816, notebook 10.

38 Adrian Stokes, 'The Sculptor Agostino di Duccio'. [Extracted from the third of four essays on the Tempio Maletestino [sic] at Rimini], in T.S. Eliot (ed.), *The Criterion 1922–1939, Vol. IX October 1929–July 1930*, London: Faber & Faber, 1967, pp. 44–60 (p. 47).

39 *CW*, 1, *The Quattro Cento*, p. 145.

40 *Works*, 12, p. 83.

41 *CW*, 1, *The Quattro Cento*, p. 142.

Chapter 3

Taste

Flora Samuel

All of life is a dispute over taste and tasting.

Friedrich Nietzsche[1]

Menu

What makes a good piece of beef? Through the close study and filming of cattle breeders in Lombardy, Christina Grasseni experiences an 'apprenticeship of the eye' and gains insight into the 'skilled jargon' and 'skilled gaze' set of values rarely articulated in words in the assessment of bovine aesthetics.[2] Similarly, Busoni, in her study of retired naval carpenters in Livorno: 'by operating *a occhio* (by sight, namely without the aid of measuring tools), the carpenter's eyesight sums up the competence derived by experience, which involves the capacity for an aesthetic and normative judgment.'[3] Yet it seems to me that, if anthropologists want to find examples of highly subjective and embodied qualitative decisions, they need look no further than the architectural crit or design review. Indeed, a close study of the profession in line with the sociologist Pierre Bourdieu's survey of the academic community, *Homo Academicus*, is long overdue.[4] The trouble is that cattle farmers are more interesting than architects. They clearly appear to be doing something useful.

Aperitif

The inspiration for this paper came from my own experience as an architect who lives in a community of non-architects – a rarity in my experience, as architects seem to cluster together. Outraged by plans for a library extension at the end of my street in Cardiff, I wrote to the council in complaint, only to be told that 'good design costs money' – something that I fundamentally refute. Bewildered by this statement, I turned to my friends and neighbours and parents at my daughters' school for support, but was further shocked by their lack of response or understanding.

They really didn't seem to care if they had an ugly, *faux* traditional and unsustainable library; something that caused me further soul searching. How could these people, my friends, not understand the importance, the life-enhancing, joyous importance of a good building? This experience brought home to me the extraordinary rift in opinion between architects and other people relative to the subject of taste. This was something that I had never really thought about before, although I now realise that there is a growing body of writing and thought on this subject – Jeremy Till's book *Architecture Depends* provides a deeply thought-provoking way into this area.[5] This paper is the result of my own attempts to come to terms with the issue of architectural taste, written from the perspective of a teacher of architectural design, and is here just in case the discussion should become too rarified.

The founding address of the Royal Institute of British Architects states that the architect has a 'responsibility to public opinion for the direction and maintenance of the national character for taste'.[6] Taste, according to the dictionary definition, is '(a manner or quality indicative of) critical judgement or discernment especially in aesthetic or social matters'. In its etymology, the word – most appropriately – links back to the senses, as it is highly subjective, having a long and complex relationship with architecture. In this paper, taste is used, much in the sense of the sociologist Pierre Bourdieu, as something that is used to confer power and status. Put crudely, an 'interest in culture' is used 'to separate the barbarians from the elect'.[7] Architects are the arbiters of taste, their position reinforced through their title, which enables them to impart 'symbolic capital' to a client in search of social distinction. In his book *Distinction: A Social Critique of the Judgement of Taste*, he tried to find out whether there was any link between social class and what people considered to be the 'ideal home', discovering that, as a person moves up the social hierarchy, there is much more interest in 'overtly aesthetic properties' such as 'imagination' and 'harmony' and much less emphasis on functional things such as being 'clean, practical' and 'easy to maintain'. Taste, then, is embroiled in issues of class, good taste being far removed from basic issues of need.[8]

Entrée

Nowhere does one feel the clash of cultures between architects and non-architects more than when interviewing prospective students for a place on an undergraduate architecture course, where applicants bring a portfolio of material to indicate their suitability for a degree in architecture. The students are invariably quite young. They may not have yet fully developed tastes of their own, but instead may be relying on those of their parents and teachers, mingled with their own, perhaps visceral, responses. Time and time again, the interviewer, if lucky, is shown images of buildings by Foster, Rogers and Gaudí. Maybe because they are most famous, but also, I would suggest, because the prospective student has not yet undergone a rigorous regime under the taste police of architectural education. This regime begins in the very first history and theory lecture that the student is subjected to when, even to this day, the 'true' history of architecture is often, in my experience, presented, unqualified by any discussion about why and how these particular exemplars were selected.[9]

This lack of critical context gives, in Bourdieu's words, 'an air of eternal youth to certain themes, although they are in reality strictly situated and dated'.[10] Here, certainly in the British context, Aalto and Le Corbusier, in their functionalist modes, reign free, and words such as 'space', 'tectonics', 'context' and 'order' remain unqualified and are only assimilated through repeated use.[11] Still lurking within all this is a belief in the existence of an ineffable and inexplicable – almost Platonic – architecture, too clever and subtle to be dissected under architectural autopsy, almost always the product of a genius architect. Rarely is the role of the client, profound in the case of the Schröder House or the Villa Schwob, for example, ever acknowledged. Tales of deeply dissatisfied clients such as Dr Edith Farnsworth are shrugged off or ignored.[12] Architecture as an autonomous art remains largely at the fore.

Much of architecture is an unwritten language that is assimilated when students enter on to an architectural course. Most architectural students are quick to learn the fact that they are unlikely to excel if they work within the classical tradition or are attracted to particular styles, no matter how much the general public might like them. As Pierre Bourdieu writes,

> the educational system never entirely fulfils its rational function: the essential part of what schools communicate is [. . .] acquired incidentally, such as the system of classification which the school system inculcates through the order in which it inculcates knowledge or through the presuppositions of its own organisation (the hierarchy of disciplines, sections, exercises etc.) or its operation (modes of assessment, rewards and punishments etc)'.[13]

Survival in the 'field' of architecture depends to a great extent on the ability to conform to the unwritten rules, or 'doxa', another term from Bourdieu, that are so prevalent in the profession.[14] No wonder architectural students get confused about the way in which they are assessed and, indeed, what architecture is all about.

Recently, in Britain and elsewhere, with the advent of Quality Assurance procedures there has been pressure on teachers of architecture to develop 'grade descriptors' (see Chapter 5), in which they set out exactly what it means to achieve, for example, that most inexplicable of grades, the 'first': the top classification in which 'talent' does invariably seem to play a role. This process invariably divides architectural educationalists into two camps.

On one side there are those who fervently believe that design is something so innate, and almost spiritual, that it defies categorisation, rules being seen, as Bourdieu points out, as 'fit only for pedants'. He who is beyond the rules 'puts himself forward as the maker of higher rules, i.e., a taste-maker, an *arbiter elegantium* whose transgressions are not mistakes but the annunciation of a new fashion, a new mode of expression, or actions which will become a model, and then modal, normal, the norm, and will call for new transgressions by those who refuse to be ranked in the mode' and so on.[15] In the words of Jeremy Till, it is the 'enigmatic quality' of architectural creativity 'that raises its value on the external market'.[16]

Those on the other side believe fervently in the importance of the descriptors as a tool to bring clarity, fairness and transparency to the process. If architectural design can be unpacked and explained, then it is something that can be learnt. It immediately becomes something more inclusive. Further, students themselves become more aware of real skills that they have developed in the course of their training; skills that are recognised and described during assessment and are, for this reason, easier to express in the marketplace. The use of grade descriptors also encourages critics to mark with some precision, enabling them to stretch the spectrum of marks used from a mealy-mouthed percentage of 72 (a borderline 'first') up into the 80s and beyond. What is the point of a Platonically perfect 100 per cent if marks of 80–100 per cent are equally unattainable?

The Architects Registration Board sets out the agenda of what an architect should learn in its 'Criteria for Assessment',[17] but it is up to the individual school to describe its grades in the field of architectural design, and so these vary to quite a large extent. Beyond, ideally, fulfilling the criteria set down for the project, everyone has experience of the project that breaks all the rules but that soars up into the stratosphere. We have all experienced the complaints of worthy and diligent students who have covered all the bases but remain at 68 per cent. Sometimes we have even been those students. This is where the hotly debated criterion of 'originality' often creeps in, and where a stunning drawing enables tutors to project on to the scheme a multitude of meanings that are rarely there. For me, the best possible projects fulfil the thesis that they set out to explore with critical rigour, depth and clarity, things that go beyond issues of taste. Time and time again I have seen that, when tutors are left without the framework of marking descriptors, differences in taste are quick to emerge.

Plat du jour

The person who, in the UK, has perhaps most applied himself to the issue of what might be considered quality in architecture is, of course, Sunand Prasad, who was quick to realise that, unless quality in design can be described, quantified and indeed celebrated, it will never appear to have any worth. It cannot be factored into building costs or be seen to bring about long-term benefits. 'Social capital' is now widely recognised in the social sciences as something that costs nothing but has huge value, economic and otherwise. Design quality should be much the same. Prasad's controversial DQIs are discussed in the next chapter. Central to the assessment of the design quality of real buildings is the practice of post-occupancy evaluation, which takes place all too rarely, largely owing to cost.

The question, then, is whether design quality in architecture schools should be the same as design quality in practice, while bearing in mind that the profession is continually complaining about the schools' inability to prepare students for the 'real world'. In his talk 'Walking the Girder', made at the Oxford Conference on Architectural Education in 2008,[18] Jeremy Till argued that the study of architecture can prepare students for practice or for something more vocational that goes beyond practice, that it challenges practice and relates to core and enduring values held by the individual. He asked whether we could trust practice to address the issues of, for example, globalisation, sustainability and social inequality. The answer, very clearly, was no. Continually pressurised by the vagaries of the marketplace, there are few practices that are prepared to expend their often-lean profit margins on addressing such issues. In Till's opinion, schools should therefore produce students who are able intellectually to address the challenges of the profession of the future, whatever it might bring. This, I would suggest, is why students should be assessed on their ability to think, research a novel situation and formulate an argument that expresses an ethical position using a variety of appropriate media. For the time being, quality in architectural practice is a rather different creature, alluding very much to the marketplace and, of course, to issues of taste.

Dessert

One person who was fully aware of these different dimensions of architecture was of course Le Corbusier, for whom vocation and practice were one and the same, although some trickery was needed to achieve this. One thing about Le Corbusier, something that is unusual in architecture unless you count Rem Koolhaas' recent ventures into glossy magazines, is a serious engagement with issues of taste. Having engineered his own meteoric ascendancy through the barriers of class, nobody has been more aware of issues of taste than he was. Indeed, his book, *The Decorative Arts of Today*,[19] is an attack on bourgeois taste, arbitrary decoration and the ill-considered use of styles. Yet he is not so dismissive of all aspects of popular culture – he celebrated women's fashion, he celebrated the filing cabinet, the suitcase and the pipe. He used furniture to cross class and gender barriers: the British club armchair is removed from London men's clubs and is repositioned in a Parisian living room, sat in by a woman who smokes and reads the paper. Believing that women have most to gain from his proposals, he targeted women's magazines and exhibitions to publicise his work. He used a mixture of the familiar and the new, pictures from popular journals coupled with his own plans and drawings to give a sensation of continuity, even to soften up his audience. He even turned his attention to nursery school education, realising the importance of instilling particular notions of taste in children at an early age in order to change the future.

Central to all this was, of course, Le Corbusier's Modulor, his system of proportion designed, to paraphrase Einstein, to make the ugly difficult and the beautiful easy. The Modulor was promulgated as a system of standardisation, yet, for Le Corbusier, it played an important educational role in making people understand the harmonies within nature. Appreciated at a gut level, Le Corbusier believed it would change people's behaviour and thinking. Appreciated at an intellectual level, it would achieve even more. Much of architecture has been built on an inherent belief in beauty, on the superiority of particular proportions and upon a dangerous conflation of meaning between aesthetics and ethics. We, as architects, can agree on good and bad proportions, but is this something that the general public can appreciate at all?

Fromage

Discerning clients employ architects for the transmission of cultural capital. The architect makes claims to aesthetic authority, and the client's own store of cultural capital is augmented through the employment of the architect. But, as I have found in my research into why people don't employ architects to design their house extensions, if people have no interest in cultural capital, they have no interest in employing an architect.[20] They can see absolutely no benefit in doing so.

It might be tempting to describe the couples that I interviewed in a Cardiff suburb during the summer of 2007 as ordinary, but half of them are part of the tiny percentage of people who, in the late 1980s, would have gone into higher education, and each couple's collective income is in excess of three times the national average. They are, in fact, relatively wealthy and well educated by British standards. Although fond of literature and pop music, none has any manifest interest in the visual arts. As such, it seems that they have no interest in 'symbolic capital', associated by Bordieu with aesthetic taste, the production of which is, in Dovey's terms, 'the architect's key market niche'.[21] I asked one couple, Rachel and David, whether, if they were shown examples of extensions designed by architects and extensions designed by builders and technicians, they thought that they would be able to tell the difference. The answer was 'probably not'. Instead, such people employ a selection of 'architectural designers', quantity surveyors, engineers, CAD technicians and others for extending their own homes. The RIBA has no official statistics on how many people do this, but the evidence is plain to see. There is no delineation between the architect, 'the architectural technologist' or indeed the 'chartered surveyor' – all are perceived equal to the task.[22] Paul Hymers, in his best-selling book *Home Conversions*,[23] describes 'a good designer' solely as 'one who possesses the necessary skills of draughtsmanship and is familiar not only with the details of construction, but also with the problems and regulations relating to the work'. At no point is there any mention of the words design or quality. The building is judged, or tasted, according to its functional qualities, value for money and ability to fit inoffensively into context, thus ensuring a good selling-on price. This brings us back to Bourdieu and his analysis of the ideal home according to class and the issue of whose taste we are talking about anyway.

Digestif

The 1836 RIBA assertion that it is the role of architects to advance the British public in issues of taste seems to be alive and well among architects. With the dissolution of belief in modernist truths, the very nature and validity of taste have been called into question. Another measure of quality is needed, to do with fitness for purpose in every sense of the word. While star architects have the power to confer mystique and cultural capital on artistic organisations in need of mandate, such things are irrelevant to the vast majority of building projects, those at the end of your street or mine that are deemed by communities and local authorities to have nothing to gain from the involvement of an architect. If aesthetic authority is all that the architectural profession really has to offer, then no wonder it has become irrelevant to the building process. The cloth must be wiped clean and the table laid again.

L'addition

1 Friedrich Nietzsche, 'On Those Who Are Sublime', in *Thus Spoke Zarathustra*, trans. Walter Kaufmann, 1883–92, 2.

2 Apparently, their favoured view of the cow is from the rear, as it enables them to assess the positive attribute of angularity, udder size and set.

3 M. Busoni, '*I ritmi della memoria. Conversazioni sui lavoro con i carpentieri navali a Limite sull'Arno*, Florence: Centro Editoriale, 1996. Cited in Cristina Grasseni, 'Video and Ethnographic Knowledge', in Sarah Pink, Lásló Kürti and Ana Isobel Afonso, *Working Images: Visual Research and Representation in Ethnography*, London: Routledge, 2004, pp.16, 28.

4 Pierre Bourdieu, *Homo Academicus*, Stanford, CA: Stanford University Press, 1990.

5 Jeremy Till, *Architecture Depends*, Cambridge, MA: MIT Press, 2009.

6 Royal British Institute of Architects, *Transactions of the RIBA*, Vol. 1/1, London: John Weale, 1836, p. vii. Quoted in Till, *Architecture Depends*, p. 154.

7 Pierre Bourdieu, *Distinction: A Social Critique of the Judgement of Taste*, trans. Richard Nice, Cambridge, MA: Harvard, 1984, p. 250.

8 Ibid., p. 73.

9 Flora Samuel, 'Lower Case History and the Development of Reflective Practice in Studio', paper presented to the Architectural Education Exchange Conference, Cardiff, 2001.

10 Bourdieu, *Distinction*, p. 73.

11 This is what makes Adrian Forty's book *Words and Buildings* (London: Thames & Hudson, 2001) so refreshing.

12 All these houses had female clients. The subject of architectural taste is closely bound up with issues of gender and the question of whether men and women design differently.

13 Bourdieu, *Distinction*, p. 67.

14 Bourdieu, *The Rules of Art*, trans. Susan Emanuel, Cambridge: Polity, 2005, p. 307.

15 Bourdieu, *Distinction*, p. 255.

16 Till, *Architecture Depends*, p. 160.

17 See www.arb.org.uk/education/arb-criteria/contents.shtml, accessed 7 June 2009.

18 Jeremy Till, Oxford Conference on Architectural Education, 24 July 2008.

19 Le Corbusier, *The Decorative Arts of Today*, London: The Architectural Press, 1987. Originally published as Le Corbusier, *L'Art décoratif d'aujourd'hui*, Paris: Editions Crès, 1925.

20 Flora Samuel, 'Suburban Self Build', *Field*, 2008, 2, pp. 111–25. Available at www.field-journal.org/index.html.

21 Kim Dovey, 'The Silent Complicity of Architecture', in Jean Hillier and Emma Rooksby (eds), *Habitus: A Sense of Place*, Ashworth: Aldershot, 2002, pp. 238–96 (288). 'Objectively and subjectively aesthetic stances adopted in matters like cosmetics, clothing or home decoration are opportunities to experience or assert one's position in social space, as a rank to be upheld or a distance to be kept.' Bourdieu, *Distinction*, p. 57.

22 In the words of 'House extension online', either of these 'consultants' will 'provide advice on what the best options are with regards meeting requirements within the constraints of the position of your property'. See www.house-extension.co.uk/planning/use_an_architect.htm.

23 Paul Hymers, *Home Conversions*, London: New Holland, 2003, p. 16.

Part 2

Production

Since the 1980s in particular, 'quality management' has burgeoned in Western societies, notably in the UK and Western Europe. Focused on efficient processes and predictable outcomes, its priorities are recorded in the European Standard ISO 9001 and the British Standard BS 5750. In public and private sectors, quality management has emphasised transparent, auditable systems that seek to streamline efficiency and attempt to manage-out potential discrepancies arising from individual actions.

This process-focused, audit-accountable management culture is often perceived to be at odds with 'fine' architecture and the architectural values of creative judgement that connoisseurs find embodied in it. As Flora Samuel argued in Chapter 3, architects are in danger of fostering the marginalisation of architectural values by indulging in a potentially introverted professional discourse about quality that can alienate others in the construction industry, building clients and society at large. There have been attempts to bring these worlds closer together, to write qualities of built architecture into management processes. In 1999, under the aegis of the UK's Construction Industry Council, Sunand Prasad co-authored 'Design Quality Indicators', a tool for use in the project briefing and scheme design process that aims to bolster built architectural quality in the power games of building procurement. Where experts in cost, time and construction management put their cases persuasively, the DQI was intended to increase the status of architectural quality of the built product – not just for architects but also on behalf of building clients and end users. Many architects, however, keen to maintain emphasis on individual judgement, rather than a judgement by consensus and negotiation, have questioned the merits of DQIs. In Chapter 4, Juliet Odgers and Flora Samuel present the DQI tool, drawing on an interview they conducted with Prasad.

Architectural education – which tends to impart design priorities through learning-by-doing in a studio culture, tested by the criticism of peer review – has also been scrutinised through the lens of quality management. In Chapter 5, Helena Webster explores attempts by the UK government's Quality Assurance Agency to introduce Subject Benchmark Statements in architectural education. Her study reveals a continuing scepticism about this process in schools of architecture, but finds merit in localised collective processes for evaluating design quality as a helpful stimulation of debate among students and teachers.

The contemporary construction industry emphasises functional efficiency in process and product. The industry's preferred modes of evaluation are primarily rational and scientific, assessing its products according to their development of on-site- and pre-fabrication systems, environmental and climatic systems and material properties, understanding exemplar projects as contributing to the development of a particular building 'type' (schools, hospitals, theatres, and so on). This pervasive ethos, which some architects find at odds with architectural culture, is in large part of architects' own making, as Adam Sharr shows in Chapter 6. It owes a debt to a scientific model of architecture that was highly influential, at least in the British profession, in the late 1960s and early 1970s.

Matthew A. Cohen, in Chapter 7, revisits construction management at Filippo Brunelleschi's thirteenth-century Basilica of San Lorenzo, relating the wildly varying workmanship, which remains apparent today, to the changing requirements of its patrons, the Medici family. To speed up construction, time and cost were increasingly emphasised, at the expense of quality. It is by no means new that a well-organised building procurement process does not necessarily produce a built outcome that has architectural quality.

Innovation – often associated with the quality appreciated in architecture by connoisseurs – inevitably involves risk. And quality management regularly prefers to reject risk where possible, seeking more control over process and more accuracy of specification. Mhairi McVicar explores these issues in Chapter 8, with reference to another church that has become part of the architectural canon: St Peter's, Klippan, Sweden, designed by Sigurd Lewerentz in the 1960s. McVicar explains that architect and contractor developed an intense working relationship in which both frequently agreed to overrule the drawings, having the nerve to reject preconceived certainties, a process that raised hugely the architectural quality of the outcome. McVicar's discussion implies that the anticipation of quality perhaps has less to do with the detail processes of production than the ethos underlying that production.

Chapter 4

Designing in quality

Juliet Odgers and Flora Samuel

What follows is based on a conversation we had with Sunand Prasad, one of the authors of the Design Quality Indicator (DQI) tool. Prasad is a practitioner with an academic background and, at the time of writing, president of the Royal Institute of British Architects.[1] *As will become apparent, the DQI was originally conceived as a way of reintroducing the conversation about the quality of the eventual building into a context dominated by concerns about procurement processes. It came to take on greater ambition and significance as an attempt to map the complexity and subtlety of architectural judgement so that it no longer remains the preserve of the initiated.*

Introduction

'Introducing the DQI process to the construction of a new building or the refurbishment of an existing building will dramatically improve the quality of the design and construction',[2] so claims the Design Quality Indicator website. The DQI process was conceived to create a common language of quality understood by all its members, the result of extensive consultation across the board, hence its slightly industrial, perhaps managerial gloss. Born out of a deep need for the acknowledgement of the benefits of good design within a culture of accountability, it facilitates the involvement of the user at every stage of the design process and, most importantly, creates a vehicle for defining the success or failure of the building itself in terms of its designed qualities as built. The DQI was designed to face squarely the forces of government and market with the intention of influencing them.

The DQI tool was conceived in 1999 by Robin Nicholson, Sunand Prasad, Michael Dickson and others to counter the dominance of process measurement through Key Performance Indicators, introduced by the Egan report and the M4I.[3] A steering group was set up within the Construction Industry Council, which had been created in 1988 through the inspiration of the eminent engineer Ted Happold, to reinforce the multidisciplinary nature of design and construction.[4] A substantial grant was received from the government, and supplemented by the Commission of Architecture and Built Environment among others, to develop a practical methodology for a shared evaluation of design quality. Jeremy Till, in his book *Architecture Depends*, sets out two competing views of architecture: one as vocation, constantly critical and shifting in response to the demands of liquid modernity; and one as practice, defined by order, market forces, rules, laws and government policy. Sunand Prasad argues that the DQI is a response to the contingent realities within which the built environment is created, a response precisely of the kind that would answer Till's critique of the profession.[5] It aims to facilitate inclusion and dialogue – to allow for the voices of as wide a constituency as possible to be included in the formation of intentions and evaluation of a building. Having described the DQI process and its origins, we will illustrate its dialogic and didactic functions, which render it more than just a simple measure of quality. In addition, we will present material from our interview with Sunand Prasad that explains the ambition of the DQI to bridge two cultures in the construction industry – on the one hand, an architectural culture of subjective value, deeply suspicious of measurement, and, on the other, a building culture that demands measurable outcomes to justify public and private investment. Prasad, in a reflection on C.P. Snow's famous 'two cultures' essay, says 'it seems that in our society it is thought that science lives by measurement and art dies if measured'.[6]

Tool

Essentially, the DQI is a tool that encourages participants in the design process to engage with issues of quality – not the quality of the construction process – but the quality of the building as a built and inhabited artefact (Figures 4.1 and 4.2). There are other measures that can be used to gauge the success of a construction process (for example, key performance indicators (KPIs));[7] equally, there are a number of tried and tested methods of measuring the sustainability of a building, both during construction and during its occupied life.[8] The DQI was not intended

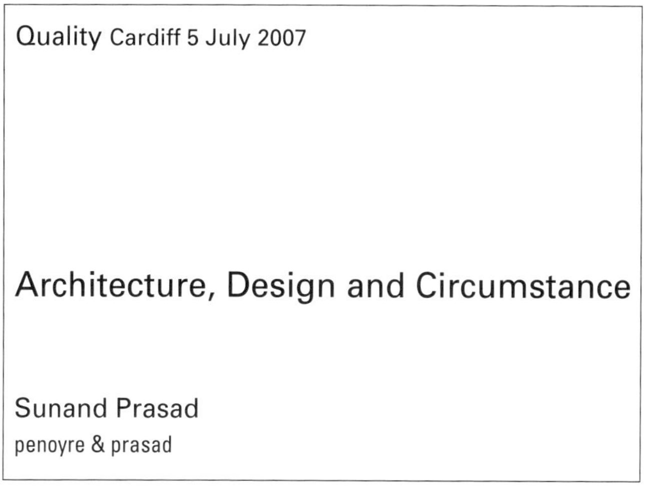

4.1
All illustrations in this chapter are from the PowerPoint presentation given by Sunand Prasad at the 'Quality' conference in Cardiff University, 2007

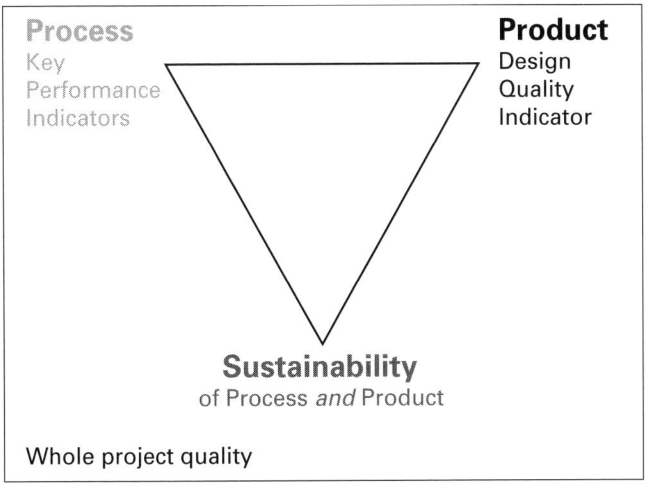

4.2
The DQI was designed to address an area ignored by performance indicators and sustainability assessment tools – the quality of the building itself, in architectural terms

to supplant these but to complement them. In fact, the concern was rather that the measures of procurement quality and sustainability should not entirely eclipse the conversation about the quality of the finished 'product'. In Sunand Prasad's words:

> This [the DQI] is entirely about product. It is not about the process by which [the building] arose, and it's not about the sustainability, whether of the process or of the product. These are very important factors but they are about circumstance. And they need their own tools and they need to be honoured in their own right. Don't mix them up; (but) this was very difficult to achieve.[9]

So how does the tool work? Essentially, it is an online tool, facilitated through workshops and meetings. The DQI is based on a series of statements, almost a hundred, grouped under a framework based on the three Vitruvian qualities of *commoditas*, *firmitas* and *venustas* (commodity, firmness and delight), translated into contemporary consumer language as functionality, build quality and impact (Figure 4.3) (we discuss the language of the tool further below). The framework headings are further elaborated under subheadings thus (Figure 4.4):

- **Functionality** breaks down to use, access, space.
- **Build quality** breaks down to performance, engineering, construction.
- **Impact** breaks down to character and innovation, form and materials, internal environment, urban and social integration.[10]

And, again, these subheadings are further articulated into the statements to which the assessment group will respond. These are supposed to include the full possible range of issues that indicate the design quality of the eventual 'product'.[11]

We will look later at how this structure was chosen, but, first, here is an outline of the process.

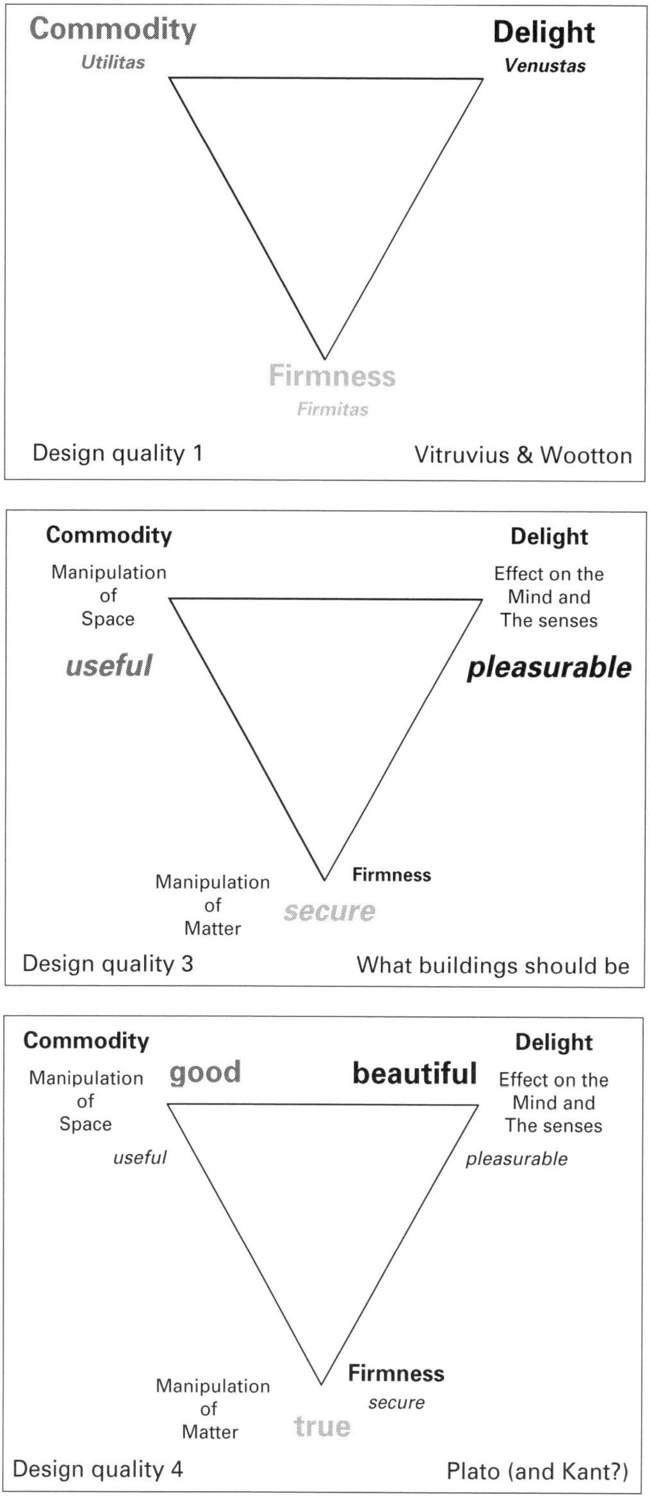

4.3
Prasad's elaboration of the framework quality fields in the language of Vitruvius and Wotton: 'commodity, firmness and delight'; 'everyday language': 'useful, pleasurable and secure'; Plato and Kant: good, beautiful and true; 'architectural discourse': 'space, meaning and tectonics'; and, finally, 'plain words': 'functionality, build quality and impact'

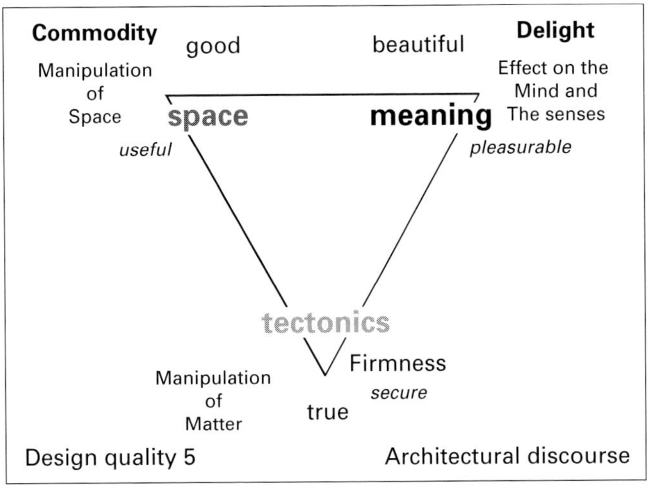

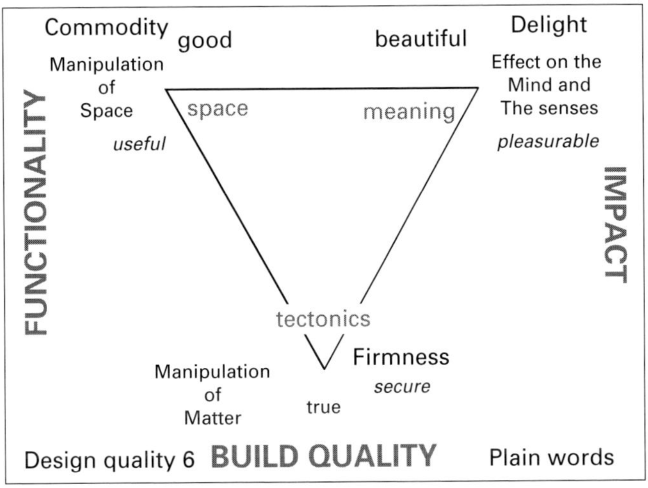

The intention is that the DQI exercise should be undertaken when the design brief is being developed. All the key people involved (representatives both of the clients and of the design teams, convened by the 'DQI leader' and guided by a facilitator) consider each of the statements from the online tool (for example 'the building should be easy to maintain') and decide on how much weight they would like to see it given in the finished building. They attribute weightings to the statements, deciding whether each is of 'fundamental' importance, or would, if achieved, give 'added value', or 'excellence' to the finished building (see Figure 4.5). A 'fundamental' aspect of the design might be essential for simple compliance, such as quantity of floor area, or daylight levels; an 'added value' aspect might be adaptability of space or integration of access with landscape design; while categories representing 'excellence' in design might involve a building that raised the spirits or brought something new to the art of architecture. The assessment group decides on its own priorities by using the tool. It also has the option of

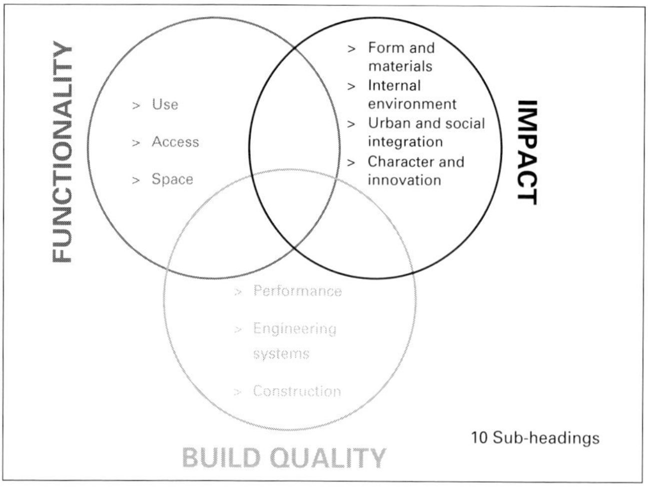

4.4
Subsections of the three quality fields that make up the framework of the DQI

deleting a statement if that attribute is not relevant. This exercise sets the weightings given to statements and aims, in Prasad's words, to 'mirror the intuitive process that any one familiar with design review uses for evaluation'.[12] Prasad explains that, to him, this categorisation is also akin to Maslow's 'hierarchy of need', which holds that it is necessary to satisfy basic functional needs before 'higher' forms of individual and collective human experience can have potency (Figure 4.6). It avoids the simple duality of 'this is more important than that', which often distorts design decisions. It recognises that a building that has high basic functionality, but no more, is unlikely to fulfil the promise of architecture, but a design that engages with architectural device and artifice without fulfilling basic functions is likely to fail. Both of these outcomes are all too common.

After the briefing exercise, the group will gather again on two or three occasions throughout the design, construction and occupation of the building to monitor the progress of the design.

The use of the tool is structured around the 'recognised phases of a building project' – preparation, design, pre-construction, construction and use – and the issues addressed at each stage, while essentially the same, are worded differently according to the stage of the process – for example, the statement from the briefing stage of the process:

'The building will accommodate the users' needs'

becomes, at the design stage,

'The building is likely to accommodate the users' needs'

and, at completion,

'The building accommodates the users needs'.

4.5
'Excellence' is achieved when the building scores highly in all three quality fields

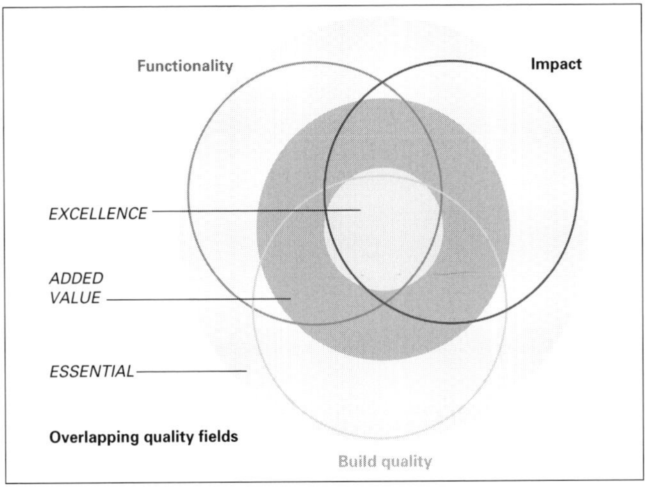

4.6
Pyramidal structure of concerns, following Maslow

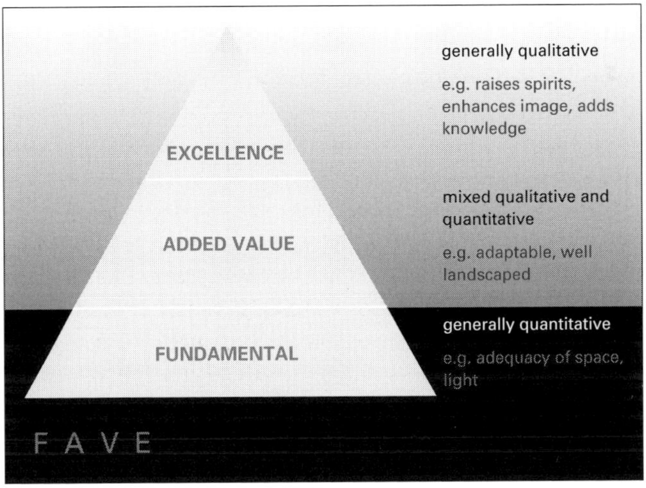

Each respondent to this 'questionnaire' scores the statements at each stage of the process according to their own judgement, marking a statement somewhere along the line from 'strongly disagree . . .' to '. . . strongly agree'.[13]

In putting together the tool, the DQI team tried to situate the wording of the questions in a mid-ground that avoided 'technical terms or jargon' on the one hand, and avoided being 'simplistic or facile' on the other. The questions were produced initially from an exercise involving the participation of 'over 40 representatives from different parts of the design and construction community' who made up the reference group of the DQI team.[14] The emphasis is on participation and easy engagement for the lay participants (Figure 4.7).

One underlying idea of the DQI tool is that the framework structures a dialogue according to the categorisations that it offers and thus facilitates a fairly rapid move to a shared understanding and meaningful exchange between client and users, and their professional advisors and contractors. As Prasad states: 'Any seminar can generate large numbers of valid but mixed-up quality headings for aspects of design. This can easily lead to a boring checklist or a set of design tick-boxes.' What the DQI tool tries to do instead is to direct the conversation in a way that will allow the participants to register their judgements accurately (clearly understanding what is objective and what is a subjective value), leading to the construction of a picture of the overall concerns of the group in a way that contributes to the emergence of a shared understanding of priorities for the building. It does this by separating out the concerns of the participants into the categories of the framework, with the clear understanding that 'excellence' is only achieved where a design scores highly in all three categories: functionality, build quality and impact (Figure 4.5).

The tool can also be used for the purposes of evaluating a building, either at the design stage or post occupancy. Both Prasad and Robin Nicholson regard this evaluative potential as less interesting than its dialogic potential. We return to a consideration of this important evaluative function below, but first – dialogue.

Development, language and intention

Finding a 'shared language' must be a priority where dialogue is the intention. 'Functionality, build quality and impact' are scarcely mainstream architectural vocabulary. So how were the words arrived at? As Prasad says:

> Looking at the hundreds of words that people came up with, it became clear firstly that we first needed to avoid category errors. For example 'value for money', 'high performance' and 'beauty' are obviously values of very different kinds. All design and construction takes place within limits of resources but that is imprecisely reflected, if at all, in the experience of the finished product. Clever design has the ability to in

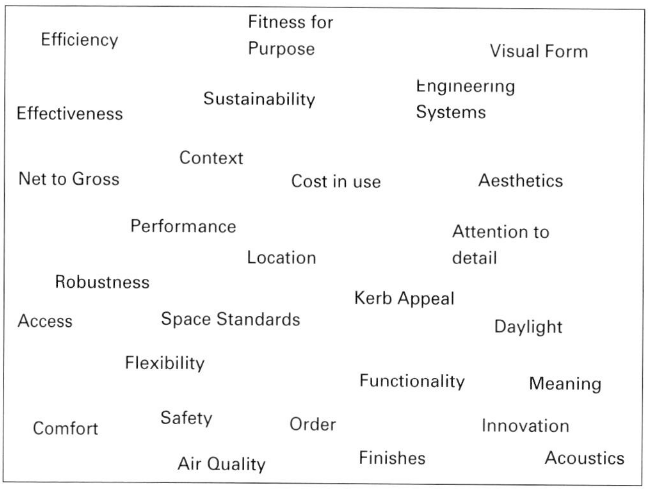

4.7
A field of concerns before ordering under the DQI framework headings

effect expand the 'resource envelope' (see Figure 4.8). The first thing we did was to make explicit the relationship between the resource envelope and the quality of the finished artefact. We then concentrated on the latter. It then became apparent that from Vitruvius onwards a three part understanding of the quality of a building (and by extension any designed artefact) has remained a meaningful way to understand our experience of designed things over a very long time. However, in their early 17th century translation 'Commodity, Firmness and Delight' carry little meaning in the development and construction industry as whole.[15]

We opted in the end for contemporary consumer language, the language of *What Car?, Which Hi Fi?*, and *Home and Garden.* When buying or selling a car people talk about 'build quality', they talk about how much luggage you can get into your car, net to gross effectively, i.e. functionality and they talk about style just what Vitruvius intended by Venustas.[16]

As mentioned above, the forty-strong reference group that helped to generate the language and structure of the DQI was drawn 'from different parts of the design and construction community'.[17] Not all were professionals, and certainly not all were architects. 'We needed new words that make sense across the

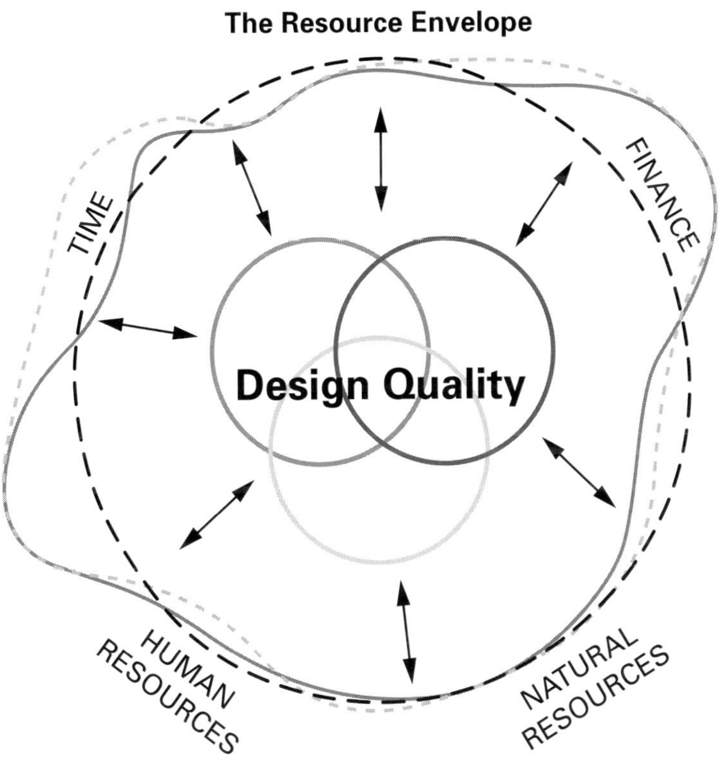

4.8
Design happens within a resource envelope

construction industry. Hence "*functionality, build quality and impact*" ', words that, according to Prasad, reflected 'various ordering structures',[18] including 'commodity, firmness and delight', from Vitruvius via Henry Wotton;[19] 'useful, secure, pleasurable', from 'everyday language'; Plato's 'good, true and beautiful'[20] – a trio of words almost painfully problematic to the current generation of architectural academics; then familiar architecture school vocabulary 'space, tectonics, meaning'.[21] And finally, '*functionality, build quality and impact*', words drawn from the journalism of consumer culture (Figure 4.3).

The domain that the DQI aims at influencing is what Prasad describes as the 'mediocre and worse-than-mediocre architecture and design in both the public and the private sector' that we all too often see in building in the UK. The PFI hospital, the building society headquarters, the affordable housing scheme – these ordinary and so often disappointing projects are the target of the DQI. 'Universalist, officially promoted tools such as quality and performance indicators cannot be aimed at encouraging genius. They are for *raising the average*.'[22] We are not talking about high-end architecture here, neither are we talking about social critique or alternative practices.

Beneath the consumer language, there lurks an intention strongly oriented towards the mores of the architectural profession. As Prasad puts it, 'what the Design Quality Indicator does is actually to describe honestly how architects think about architectural quality. That's what it does.'[23] This is perhaps not surprising if we consider that, again in Prasad's words:

> The first outside people I tried it out on were architects involved in judging RIBA awards: such as Amanda Levete, Robert Adam, Jeremy Dixon and Edward Jones . . . I was asking 'we make these judgments all the time. Can we actually write down how we do it?'[24]

On the word *impact* Prasad has the following observations:

> Sunand Prasad: Yes, *impact* – horrible word but . . .
>
> Juliet Odgers: Well I am quite interested in it because it obviously came from somewhere.
>
> Sunand Prasad: We tried everything. Nobody could think of a better word. 'Delight' is a weak word. Because 'delight' doesn't capture the intellectual intent very much.
>
> And 'beautiful' is a wonderful word, but it doesn't capture a value really important to all of us: the relationship of the work to architectural culture and the contribution it makes in the generality of architectural culture. In fact that's what we're really interested in when we, as architects, look at buildings.
>
> [. . .] So, for example, why did Zumthor's Baths at Vals become so talked about? . . . It was as if something had been re-found that had been lost in architectural discourse. And suddenly we thought 'oh yes, we forgot about that'. That building then makes a contribution,

> it becomes an anchor in the language. So we can think of a number of buildings like that. I guess Foster's Willis, Faber, Dumas was a major event in architectural culture. To a lesser degree perhaps Cullinan's Downland Grid Shell; and certainly Frei Otto's Munich Stadium and Utzon's Sydney Opera House. And of course Corb's Ronchamp and much else [. . .] The biggest thing about these buildings isn't their functionality or their build quality, it's actually [that] they change the way we see. I don't think that 'beautiful' and 'delight' embrace that essence. Whereas 'impact' does. Because 'impact' – weird word though it is – at a very prosaic level it is simply 'how does it affect you' and 'how does it affect the surroundings'? What impact does it have on the neighbourhood? What impact does it have on the user walking in? Impact is about what a building does to you and then: what is the impact of the building on the world? What is the impact of the building on knowledge? Does it contribute to new knowledge? That's impact.
>
> If you look at the subheadings and the questions underneath 'Impact', you [. . .] find a question 'the Design has [. . .]advanced the knowledge about design and buildings'. And that's what is common to all those examples which I have just mentioned, they have advanced our knowledge. And that's a magical property of a work of art [. . .] it can have this big effect outside the project. And of course no project manager and not all clients or even engineers in the group were focused on that, whereas architects are very passionate about that.[25]

So despite reservations about the language of the DQI tool – 'I don't like the word DQI – it's awful' – Prasad in the end embraces that language and expresses a modest triumph in embedding what he sees as core architectural values in the tool. As he explains its process of development, the tool comes out as the end product of a process of translation – 'how architects think about architectural quality', translated into 'words that make sense across the construction industry', a series of 'Plain English' statements, mapping values from the objective to the subjective:

- The building is accessible to wheelchair users.
- The building will be easy to maintain.
- The air quality will be appropriate to its use.
- The building will contribute to knowledge of design and building.
- The building will make you think.

So, in attempting to raise the level of average, what the architects behind the DQI are doing is explaining what they value in design and encapsulating it in an accessible language and structure. There is a dialogic intent in this – there is also, *sotto voce*, a didactic intent, which we will discuss below.

The tool is only as good as its users – dialogue or measurement

> When you are fighting against big cultural currents these things (i.e. the DQI) have quite weak traction. I can argue passionately for its value and at the time everyone will agree. But often when someone in the audience tries later to explain it to somebody else they can't capture it. That's a big test of whether you've actually communicated an idea [. . .] whether it has third party traction. And I have seen over and over again [. . .] it will tend to degrade to a different level.[26]

Prasad is well aware that a DQI tool is only as good as those who use it, and, in practice, it has been found to be the case that good facilitation is essential if the hoped-for inclusive dialogue is to be achieved. Good facilitation is necessary for a successful process. Similarly, both Prasad and his co-author of the DQI, Robin Nicholson, believe the dialogic potentials of the tool are more interesting and useful than the evaluative potentials.[27] The moment the use of the tool degrades from a rounded understanding, emerging from conversation and consultation, into a tick-box approach, the core of its potential is dissipated. Try as you might, it is difficult to find a way in which passionate engagement and expert knowledge, wielded in the service of a group intention, can be *superseded* by any tool.

Prasad's argument for using and continuing to develop the DQI rests partly on the thought that, if we are to engender an improvement in the middle ground of buildings, to bridge the gap between the 'mediocre and worse-than-mediocre' and 'isolated works of brilliance', we must do so by increasing 'the level of critical awareness amongst commissioning clients, the professions and the industry'. The DQI then has an important didactic function:

> If the project manager actually says 'does this building raise your spirits when you come in?' then I am quite glad .The didactic impact of being exposed to such a possibility over and over again: people start to say, 'it's true actually [this building does (or does not) raise your spirits when you come in].[28]

Prasad sees a reluctance on the part of architects to engage with the predominant culture of accountability and measurement as part of the problem that he is attempting to address. We see this preoccupation reflected in his insistence that the DQI 'actually describe[s] honestly how architects think about architectural quality'.[29]

'I find it very irritating that architects with their intellect aren't actually in there trying to persuade people who think differently from them. I sometimes think architects don't really want to communicate.'[30] While the gaze of the profession is fixed on the high summits of 'isolated works of genius', or 'self referential academism',[31] the middle ground slips further and further from their influence. 'There appears to be no evidence that, by itself, the appearance of buildings highly acclaimed by architectural cognoscenti does much for encouraging quality and clients' aspirations generally,' he argues – 'what is needed is to reach in a direct way those commissioning buildings and provide them with a means to raise their game.'[32] The DQI is his proposed means. This is, of course, a response to a particular historical situation, the 'accountability culture' of late 1990s Britain, in which measure becomes an inescapable adjunct of judgement, also typically bearing an intimate relationship with public and institutional funding and commissioning. Speaking of this time, Sunand Prasad says: 'We were alarmed by what was happening in the construction industry and the fact that the process boys were just taking over, M4I and KPI were going to dominate.'[33] The experts in cost, time and construction management were speaking their concerns persuasively and loudly, and the voices that speak for the quality of the end product were consequently becoming increasingly difficult to hear and little heeded. Where only efficiency of process is measured or seen as measurable, the quality of the eventual product – the quality of the building – can be almost entirely discounted. Not measurable, therefore not measured, discounted, irrelevant, excluded. This is what the instigators of the DQI were so scared of. The DQI 'represents a constructive first step'[34] for use in 'messy, pressured, live project conditions'.[35] Its fundamental aim is to bring a common language and transparency into the discussion of design quality as a part of a complex power struggle which is architecture. As Sunand Prasad says, 'The decision to devise DQIs . . . [was] . . . an act of appropriation . . . *Measurement is too important to be left to the measurers.*'[36]

Notes

1 Sunand Prasad is principle of Penoyre & Prasad LLP.

2 www. dqi.org.uk/website/buildings/process (accessed 16 March 2009).

3 For the Egan report, see Department of Trade and Industry, *Rethinking Construction*, London: HMSO, 1998; 'The Movement for Innovation (M4I) was formed in November 1998 to implement, across the whole of the industry, the recommendations contained in "Rethinking Construction". The Movement aims to lead radical improvement in construction in value for money, profitability, reliability and respect for people, through demonstration of best practice and innovation.' Available at www.constructingexcellence.org.uk/resources/az/view.jsp?id=290 (accessed 16 June 2009).

4 D.M. Gann, A.J. Salter and J.K. Whyte, 'The Design Quality Indicator as a Tool for Thinking', *Building Research and Information*, vol. 31, no. 5, pp. 318–33. D.M. Gann and J.K. Whyte, 'Performance Indicators for the Design of Buildings', Final Report, Ref. No CIC00003, DETR, London, 2003.

5 Authors' interview with Sunand Prasad, 9 February 2009.

6 C.P. Snow, *The Two Cultures and the Scientific Revolution*, Cambridge: Cambridge University Press, 1959; Prasad, Interview.

7 For KPIs sponsored by the Construction Industry Council, see Frans Henderieck, 'Building and construction process innovation: A framework for measuring innovation performance'. Available at www.research.scpm.salford.ac.uk/.../Building%20and%20construction%20process%20innovation.doc (accessed 1 September 2009).

8 Sunand Prasad, 'Inclusive Maps', in Sebastian Macmillan (ed.), *Designing Better Buildings: Quality and Value in the Built Environment*, London and New York: Spon, 2004, p. 179.

9 Prasad, Interview. For a critique of the design of the DQI from within the culture of evaluation and questionnaire design see Sunand Prasad 'Clarifying Intentions: The Design Quality Indicator', *Building Research and Information*, Nov–Dec 2004, vol. 32, no. 6, pp. 548–51; T.A. Markus, 'Lessons from the Design Quality Indicator', *Building Research and Information*, vol. 31, no. 5, 2003, pp. 399–405; G. Dewluf and J. Van Meel, 'Sense and Nonsense in Measuring Design Quality', *Building Research and Information*, vol. 32, no. 3, 2004, pp. 247–50.

10 The structure of the DQI is intended to evolve over time. The breakdown of categories given here is from Sunand Prasad's presentation at the Quality Conference at the Welsh School of

Architecture, in July 2007. See also Whyte, Gann and Salter, 'Building Indicators of Design Quality', in Macmillan (ed.), *Designing Better Buildings*, pp. 198–9.

11 For a full description and history of the tool by its originators see *Designing Better Buildings*, particularly Chapters 15–17, pp. 175–205.

12 Prasad, Interview.

13 Prasad presentation Quality Conference, 2007.

14 Whyte, Gann and Salter, 'Building Indicators of Design Quality', in Macmillan (ed.), *Designing Better Buildings*, p. 198.

15 Henry Wotton, *The Elements of Architecture* (1624) was the first translation (free translation) into English of Vitruvius' *De Arcitectura*, usually translated as *The Ten Books on Architecture*.

16 Prasad, Interview.

17 Whyte, Gann and Salter, 'Building Indicators of Design Quality', in Macmillan (ed.), *Designing Better Buildings*, p. 198.

18 Prasad, Interview.

19 Henry Wotton, *The Elements of Architecture*, 1624, London.

20 Prasad relates 'good, beautiful and true' to Plato and Kant; see Prasad 'Inclusive Maps', p. 180.

21 Prasad, 'Inclusive Maps', p. 179.

22 Prasad, 'Inclusive Maps', p. 178.

23 Prasad, Interview.

24 Prasad, Interview. Amanda Levette, founding partner of Future Systems, now principle of Amanda Levete Architects; Robert Adam, principle of Robert Adam Architects; Jeremy Dixon and Edward Jones, principles of Dixon Jones.

25 Prasad, Interview.

26 Prasad, Interview.

27 Prasad, Interview, and Nicholson personal communication, May 2009.

28 Prasad, Interview.

29 Prasad, Interview.

30 Prasad, Interview.

31 Prasad, Interview.

32 Prasad, 'Inclusive Maps', p. 178.

33 Prasad, Interview. For the context in which the DQI tool was developed see also *Designing Better Buildings*.

34 Sunand Prasad, 'Clarifying Intentions: The Design Quality Indicator', *Building Research and Information*, vol. 32, no. 6, pp. 548–51 (p. 550).

35 Prasad, Interview.

36 Prasad, 'Inclusive Maps', p. 176.

Chapter 5

Assessing design quality in architectural education

Helena Webster

One consequence of the extensive twentieth-century literature on the production of cultural values[1] is that it is now generally accepted that notions of 'design quality' are constructed and contested within fields of practice, including architecture. It is also accepted that academic institutions play an important reflexive role for fields by offering diverse exploratory environments through which established notions of design quality can be learnt, challenged, subverted or re-invented. However, the notion that heterogeneity is an important and necessary condition for a dynamic field appears to run counter to the post-Dearing culture of parity and accountability that pervades higher education in the United Kingdom today. Increasingly, schools of architecture are required to respond to external demands to provide evidence that they define and assess design according to predetermined benchmarks of quality. This paper reports on the findings of a small-scale qualitative study that investigated the ways in which seven schools of architecture were responding to these demands.

Context

The 1997 *Dearing Report*[2] recommended wide-ranging changes to the delivery of higher education in the United Kingdom. The government subsequently founded the Quality Assurance Agency (QAA) to realise many of the report's ninety-three recommendations. Two of the QAA's initiatives specifically addressed notions of quality. First, the Subject Benchmark Statements[3] claimed to: 'provide a means for the academic community to describe the nature and characteristics of programmes in a specific subject [and] represent general expectations about the standards for the award of qualifications.'[4] Second, the *QAA Code of Practice on Assessment*[5] presented a range of precepts aimed to help institutions to carry out the assessment of students in ways that were valid, reliable, fair and transparent. In response to the publication of these documents, most higher education institutions set about revalidating their programmes, both to align with the subject benchmarks and producing learning, teaching and assessment strategies that aligned with the code of practice. However, the subsequent process of embedding these high-level initiatives within disciplinary contexts has not been as simple as the QAA hoped. Perhaps most surprisingly, the acceptance of the notions of parity and transparency in relation to the assessment of design has proved particularly problematic for architectural education, a field that traditionally values heterogeneity, tacit intuition and connoisseurship.[6]

The study

Anyone who has been involved in the assessment of design in architectural schools over the last few years will have probably noticed that academics have been either reluctant to, or struggled to, conform to the QAA's new quality requirements. However, a literature search reveals that no one has investigated the nature or extent of these problems, although Allan Davies's and Barry Jackson's research identifies similar issues in art and design.[7] Thus, it seemed apposite to find out, through a qualitative study, how schools of architecture were negotiating the QAA's new quality demands in relation to design, and why there were so many apparent difficulties.

In the first thee months of 2006, seven schools of architecture[8] were visited, and semi-structured interviews were carried out with the members of staff responsible for overseeing teaching, learning and assessment in their schools; they were all, incidentally, also design tutors. The participants were invited to respond to a series of twenty-two pre-planned open questions in hour-long interviews[9] that were designed to explore a number of key themes:

- institutional learning, teaching and assessment policies;
- national benchmarks defining design quality in learning outcomes;
- assessment policies and defining design quality for assessment purposes;
- assessment procedures for assessing design quality;
- processes for evaluation and review.

After the interviews had taken place, the responses, together with a number of supporting documents (Table 5.1), were collated and then analysed against the key themes.

Table 5.1 Supporting study documents

General documents	School-specific documents
QAA Benchmarks Statement: Architecture (2000)	Learning, teaching, assessment policies (incl: university/faculty/ school/ department/ programme)
QAA Code of Practice Section 6: Assessing Students (2000)	Programme handbook
QAA Code of Practice Section 4: External Examiners (2000)	Design course/module descriptor/handbook
RIBA Criteria for Validation (2003)	Design course briefs
ARB Criteria of Assessment (2003)	Formative and summative design assessment forms

Discussion of findings

National benchmarks defining design and design quality through learning outcomes

The QAA Subject Benchmark Statement for undergraduate architecture programmes, which broadly aligns with the professional validation criteria,[10] purports to describe 'the nature and characteristics of programmes in architecture' and the 'general expectations about the [threshold] standards for the award of qualifications'.[11] The benchmark statement's description of a graduate's (threshold) ability in relation to design is detailed in Table 5.2. One of the requirements for both university and professional validation is a 'mapping' that demonstrates the way in which programmes ensure that all students attain all the learning outcomes at the minimum threshold level. The interviews with the seven participants revealed that, while schools dutifully carried out the mapping exercise, many thought that the notion of national benchmarking itself was 'over-constraining' and 'restricted diversity'. Additionally, when asked about the validity of the benchmark statements in relation to design, none of the interviewees believed that the design learning outcomes adequately described the characteristics of a graduate designer.

Table 5.2 QAA Subject benchmark – architecture: threshold design learning outcomes

The student will demonstrate:

- an ability to produce an architectural design;
- an understanding of the ways that the analysis, research, development and preparation of an architectural programme or brief can inform design;
- knowledge of architectural histories and theories, physical and cultural contexts, and of the ways that they can inform design processes, programmes and proposals;
- knowledge and understanding of the principles of building technologies in relation to human needs;
- an understanding of the impact on architectural design of regulatory frameworks, the needs and aspirations of clients and building users, and the requirements of the wider community.

The interviewee from school C pointed out the inadequacy of the statement: 'an ability to produce an architectural design' as a threshold learning outcome. She suggested that, on a conceptual level, the ability to design should be presented as the core, or high-level, learning outcome, rather than just one of many, and that 'an ability to produce an architectural design' was not an adequate description of a threshold level. Clearly, she remarked, 'one could design well or badly'. In all cases, the interviewees believed that, because the benchmarks contained inadequate descriptions of design or design quality (even at a threshold level), schools continued to make judgements of design quality through the tacit agreement of assessors.

However, the interviewees were more positive about the QAA's 'concept' of defining and assessing a student's design ability *via* learning outcomes.[12] The new institutional requirement to use learning outcomes resulted in what they saw as a positive shift away from product-oriented assessment *via* final design drawings in Crits[13] (Figure 5.1) and towards process-oriented assessment *via* portfolios in Portfolio Reviews (Figure 5.2).[14]

Thus, despite attempts by the QAA and the professional bodies to regulate the quality of graduates through explicit statements of quality and standards, those interviewed expressed little enthusiasm for the concept of national benchmarking and little confidence in the validity of the current architecture benchmark, particularly in relation to design.

Teaching, learning and assessment policies: defining design and design quality

Given the government's financial incentives for institutions to produce learning, teaching and assessment strategies,[15] it is not particularly surprising that the seven

5.1
The 'Crit' – the unreliable and now largely superseded means for assessing design in the UK

5.2
The 'Portfolio Review' – the dominant contemporary method for assessing design in the UK

case-study institutions had university-level policies that aligned closely with the *QAA Code of Practice on Assessment.* All university policies required academic programmes to have explicit learning outcomes, assessment criteria[16] and indicators of quality.

However, given that only two out of the seven case-study schools had school-specific learning, teaching and assessment policies, in reality, module leaders in five schools had to work individually to interpret high-level generic policy into their specific subject contexts. The interviewees from these schools suggested that this was proving to be a tall order, particularly in the area of design. Yet, an initial collation of the research documents suggested that all schools had made considerable progress towards meeting the new demands (Table 5.3). For instance, all validated design module descriptions contained statements of learning outcomes that mapped on to the benchmark statements (although the quality of these learning outcomes varied widely). Design module learning outcomes consistently appeared again in module handbooks and/or project briefs, where, in four schools, there were also explicit assessment criteria (although the link between learning outcomes and assessment criteria was not always clear or valid). Additionally, notions of different levels of design quality were explicitly reified through general level descriptors in three of the seven schools (see Table 5.4), although these descriptors were located in handbooks and therefore remote from the point of use.

In three schools, design quality descriptors were indicated on assessment forms against either learning outcomes (Table 5.5) or assessment criteria (Table 5.6), albeit with varying degrees of sophistication. In addition, all but one school had designed specific assessment forms to assist the validity, reliability and

Table 5.3 Reification of design quality in programme documents

	School						
Type/location of design quality reification	**A**	**B**	**C**	**D**	**E**	**F**	**G**
Handbooks (programme/module)							
Learning outcomes (LOs) in handbooks	X	X	X	X	X	X	X
Grade descriptors in handbooks	X			X	X		
Assessment criteria in handbooks (weighted)							
Assessment criteria in project briefs (unweighted)			X		X		
Design project briefs							
LOs in project briefs	X				X		X
Grade descriptors in project briefs							
Assessment criteria in project briefs (unweighted)							
Assessment criteria in project briefs (weighted)	X						
Design assessment forms							
LOs		X		X		X	
Grade descriptors							
Assessment criteria (weighted)	X						X
Assessment criteria (unweighted)			X			X	
Generalised LO or criteria-based level descriptors				X			X
Specific LO or criteria-based level descriptors			X			X	
Written feedback	X	X	X	X		X	X

Table 5.4 School A: generic undergraduate level descriptors

1st: 79% or over	Exceptionally outstanding work
1st: 70–80%	Outstanding in most required aspects of work. All aspects of work given attention, no significant omissions or flaws, well worked out in detail.
2.i: 60–69%	Excellent in some required aspects of work. At least competent in all required aspects of work. Few or no significant omissions or flaws; well worked out in detail.
2.ii: 50–59%	Average overall standard in all required aspects of work – generally competent. May show little enthusiasm or commitment. Not necessarily fully worked out.
3: 41–49%	Below-average standard in some required aspects of work. Competent. Dull work, showing little enthusiasm or commitment. Not necessarily fully worked out in detail.
Pass: 40%	Significantly below-average standard in some required aspects of work. Some omissions or flaws. Some signs of lack of ability or application. Little attention to detail.
Fail: 31–39%	Unsatisfactory work. Serious omissions or flaws. Shows little sign of understanding. No attention to detail.

Table 5.5 Third-year-undergraduate learning outcome-based assessment form

Learning outcomes	**Levels of achievement**				
An ability to:	**Very good**	**Good**	**Fair**	**Pass**	**Fail**
Define your individual brief through drawn representations of observed processes, physical and cultural conditions			X		
Assess and prioritise architectural concerns and derive consequent architectural strategies, represented visually	X				
Explore the specific context of the site and its relationship with the design proposal through various media		X			
Communicate design processes through an appropriate use of sketches, drawings, collage, modelling, photographic, video or digital techniques	X				
Make coherent architectural proposals, which implement a degree of professional skill and judgement		X			
Design appropriate technological proposals through detailed resolution of your project and its landscape		X			

transparency of the assessment process, and these forms were in use for interim as well as final assessments (Table 5.3). Although the information on these assessment forms varied, all six contained either learning outcomes or assessment criteria, four allowed space for indications of quality, and all six had space for written comments (Table 5.7).

Although the above descriptions make it appear that the seven case-study schools had made significant progress towards meeting the QAA's requirements, further analysis of the documents, together with the interview responses, revealed that design tutors were wrestling with some significant conceptual problems.

First, there appeared to be questions about the validity of the stated design learning outcomes. It emerged that learning outcomes were usually written by one person and often repeated, for reasons of expediency, the external benchmark learning outcomes. Thus, the general lack of support for the efficacy of the benchmark learning outcomes, discussed above, resulted in a lack of support for design module learning outcomes. The interviewees reported that module leaders who attempted to write their own learning outcomes often found the language of learning outcomes complex and alien. As a consequence, design tutors often ended up variously: assessing student work against learning outcomes that they did not value; completely ignoring the learning outcomes when assessing student work; or deriving their own criteria of assessment (Tables 5.6 and 5.7) that reflected the design process, but did not necessarily align with the module learning outcomes.

Second, the interview results suggested that many design tutors had fundamental objections to any attempts to reify notions of good design or what

Table 5.6 School C: diploma criteria-based assessment form (with quality indicators against criteria)

Assessment criteria	**Level of performance**				
	Best	**Good**	**Fair**	**Adequate = ARB/RIBA criteria**	**Unacceptable**
1. Analysis and definition of an architectural brief/ problem/context	Evidence of highly individual analysis, evaluation and synthesis that demonstrate a deep understanding of the brief/ problem and clearly inform its development	Evidence of consciously selective analysis, evaluation and synthesis that clearly inform the development of the brief/problem	Evidence of a broad range of relevant information and some evidence of analysis that clearly informs the development of the brief/ problem	Evidence of a broad range of relevant information but lacks analysis that clearly informs the development of the brief/ problem	Evidence of a limited collection of random information that does not obviously inform the development of the brief/ problem
2. Generation and evaluation of architectural ideas/concepts	Demonstrates high-level critical/ reflective thinking that extends architectural ideas and forms to inform the exploration of original + innovative architectural concepts	Demonstrates extensive critical and reflective thinking that draws on existing architectural ideas and forms to inform the exploration of architectural concepts	Demonstrates some critical and reflective thinking that draws on existing architectural ideas + forms to inform the exploration of architectural concepts	Demonstrates the evaluation and synthesis of relevant formal + technical knowledge that generates simple architectural concepts	Demonstrates a limited evaluation or synthesis of relevant knowledge that fails to generate plausible architectural concepts
3. Development of detailed formal architectural and technical solution(s)	Evidence of highly original research, evaluation and synthesis that inform the development and formal/technical resolution of original architectural designs	Evidence of extensive research, evaluation and synthesis of relevant formal + technical areas that clearly inform design decisions and result in coherent architectural designs	Evidence of some relevant theoretical/technical research and its application towards the development and resolution of coherent architectural designs	Evidence of the integration of standard technical and formal principles towards the production of routine architectural designs	Evidence of the partial/limited integration of knowledge of the standard technical and formal principles towards the production of a routine architectural design
4. Communicate ideas, processes and detailed solutions	Selects and uses media in a highly directed and original way to communicate intended architectural form and meaning	Uses a wide range of media to comprehensively express architectural ideas and designs	Uses a range of techniques and media to successfully express architectural ideas and designs – some elements missing or unclear	Uses a limited range of techniques and/or media to communicate routine ideas and designs – some elements missing or unclear	Uses a range of techniques and/or media that falls short of communicating coherent ideas or designs

Table 5.7 School A: First-year weighted criteria-based assessment form (most design tutors objected to weighting on the grounds that it makes assessment too rigid/inflexible)

Final reviews		**Date**
Student name	**Mark**	
Integrated design 25%		Comments
Response to user needs 15%		
Response to site 15%		
Tectonics 15%		
Process 15%		
Any missing drawings		
Task for coming week		

might constitute a good designer. Indeed, they often vehemently resented the external imposition of what they saw as an alien, pseudo-scientific, 'objective' mindset on to a cultural discipline in which the most important values were qualitative – for example, 'design' and 'artistry'. In their minds, notions of design quality were 'tacitly' held by connoisseurs, culturally constructed and constantly open to contestation and revision, and, as a consequence, they believed that design quality could not (and should not) be reified in any ahistorical sense.

Finally, the interviewees reported that the majority of design tutors remained to be convinced that the existing 'hidden' processes of assessment through 'connoisseurship', which included marking by design tutors, cross-design team moderations and intensive ratification of standards by external examiners, were perfectly valid, reliable and fair. They could not understand the point of generating more paperwork just to make an (inadequate) record of the existing robust procedures. Thus, although the paperwork existed to placate the external agencies, in reality, it appeared that the reified notions of design and design quality that were enshrined in design module documents and assessment forms were largely perceived by design tutors as essentially 'inadequate' and 'misguided', but a necessary evil.

Yet, interviewees also cited some more positive experiences. The interviewee from school C, who had looked at the generic learning and teaching literature on assessment,[17] explained that she had formulated the quality descriptors (Table 5.6) for her school's Diploma design assessment form by gathering design tutors around student portfolios and asking them, first, to describe the criteria against which they assessed the portfolios and, second, how they differentiated between poor and excellent performance under each criterion. Interestingly, she reported that the design tutors found themselves agreeing on the core criteria, which, incidentally, mirrored the design process, but not on their relative importance. In the end, the design tutors agreed that the criteria should be left unweighted, thus allowing scope for a synoptic view to influence the final mark.

Conclusions

Seven years have passed since the introduction of national subject benchmarks and requirements for higher education institutions to assess students against explicit statements of quality. So, how have schools of architecture fared in relation to defining and assessing design quality? Clearly, the QAA's well-intentioned idea that disciplines could agree on national benchmarks, including statements of scope and threshold quality, has turned out to be more problematic than envisaged. The results of this study suggest that there was little support in schools for either the idea of a national benchmark statement or its current form.[18] Additionally, the study found that the requirement to measure student performance against explicit quality descriptors was also proving unpopular and problematic for the design tutors. The study found that, although many schools were making explicit quality descriptors, design module leaders experienced significant conceptual and practical difficulties in establishing truly valid statements of learning outcomes, assessment criteria and levels of achievement. These difficulties were found to be partly a result of module leaders having to learn 'education speak'; partly because of the intrinsic difficulties attached to reifying complex tacit values; and partly because module leaders largely worked alone rather than collaboratively. Additionally, the naturalised connoisseur model for assessing design quality meant that those charged with the task of reification were often working in a culture of suspicion.[19]

So, we are left wondering if the QAA's project is worth a candle? The findings of this research suggest that the drive to harmonise quality and standards nationally has been perceived by architecture schools as constraining diversity and unnecessarily bureaucratic. It has also caused individual design module leaders in schools to devise procedures for defining and assessing design quality, the results of which seem to have suspect validity. Yet, in the two schools where design tutors worked together from first principles to identify and then reify criteria for defining design quality, the results were perceived, at least by the tutors, as valid. However, one might argue that, even in these cases, the reification of tutors' tacit notions of quality did not substantially increase the validity, reliability or fairness of the assessment process. So, what is the point? It might be argued that the point is that those who will benefit from the reification of tacit values are the students. First, students might begin to understand that notions of design quality are not purely subjective, but rather, following Bourdieu's notions of fields and capitals, that groups construct notions of design quality in particular contexts (in this case, a group of design tutors as a sub-group of the field of architecture). Second, as a consequence of the first point, once students understand what is expected of them, they should be better equipped to assess and then direct their own learning. Thus, the arguments for transparency are less to do with improving the validity and reliability of design assessment and more to do with empowering students to direct their own learning.

Finally, although the results of this case study suggest that there is a justified scepticism in schools of architecture about the efficacy of national benchmarks of quality, there are persuasive arguments for promoting a localised, collective process of generating quality statements that might stimulate a new and much needed reflexive debate about how notions of quality are constructed, challenged and redefined.[20]

Notes

1 See Pierre Bourdieu, *The Field of Cultural Production*, Cambridge: Polity Press, 1993; Michel Foucault, *The Archaeology of Knowledge*, trans. A.M. Sheridan Smith, London: Routledge, 1972; Raymond Williams, *Culture and Society 1780–1950*, London: Penguin, 1966.

2 The government appointed a committee in 1996 to make recommendations for the future of the higher education sector. The committee made ninety-three wide-ranging recommendations, including the formation of the QAA to oversee the reform of quality and standards. See Sir Ron Dearing, *The Dearing Report: Higher Education in the Learning Society. Report of the National Committee of Inquiry into Higher Education*, Hayes, Middlesex: NCIHE Publications, 1997.

3 QAA Subject Benchmark Statements responded to Recommendation 25 of the *Dearing Report*. The architecture benchmark relating to undergraduate degrees was published in 2000.

4 Quality Assurance Agency, *QAA Benchmark Statement: Architecture*, Gloucester: QAA, 2000, p. 0.

5 Quality Assurance Agency, *Code of Practice for the Assurance of Academic Quality and Standards in Higher Education*. Section 6: Assessment of Students, Gloucester: QAA, 2000.

6 Eisner defined connoisseurship as 'the art of [expert] appreciation', a tacit or intuitive skill that is learnt through experience. See Elliot W. Eisner, *The Art of Educational Evaluation*, London and Philadelphia: Falmer, 1985, pp. 87–119.

7 See Allan Davies, 'Assessment and Transferable Skills in Art and Design', Deliberations Educational Database, available at www.city.londonmet.ac.uk/deliberations/Subjects/art_design/davies_res.html (accessed 5 June 2007); Barry Jackson 'Assessment Practices in Art and Design: A Contribution to Student Learning?', in Graham Gibbs (ed.), *Improving Student Learning – Through Assessment and Evaluation*, Oxford: Oxford Centre for Staff Development, 1995.

8 The seven case-study schools of architecture included three pre-1992 universities and four post-1992 universities. Six schools were in England, and one was in Scotland.

9 The questions were loosely based on those described in SACWAG (2005) 'Developing Assessment Criteria: A Small Case Study', in Geoff Timmins *et al.* (eds), *Teaching and Learning History*, London: Sage, 2005, pp. 185–91.

10 See Architects Registration Board (ARB), *Prescription of Qualifications: ARB Criteria*, London: ARB, 2003; and RIBA, *Criteria for Validation*, London: RIBA, 2003.

11 Quality Assurance Agency, *Subject Benchmark Statement Architecture*, Gloucester: Quality Assurance Agency for Higher Education, 2000.

12 A learning outcome defines what a student should be able to demonstrate, in terms of knowledge, understanding or ability, at the end of a module.

13 The 'Crit' is traditionally a ritual in which students present their designs to a panel of assessors, with a student audience.

14 'Portfolio Reviews' are usually private events in which a panel of assessors review student portfolios against explicit learning outcomes or assessment criteria.

15 The Teaching Quality Enhancement Fund *Circular 02/24* announced a fund of £31 million per annum, for the three years from 2002 to 2005, that would be allocated to institutions who submitted revised learning and teaching strategies and action plans.

16 'Assessment criteria' define the characteristics that a piece of work must demonstrate to meet one or more learning outcomes.

17 For an overview of the literature, see Lee Dunn, *The Student Assessment Handbook*, London: Routledge Falmer, 2004.

18 The QAA is currently circulating a draft revision to the *Benchmark Statement for Architecture* that aligns more closely with the *EU Directive*.

19 For arguments in support of the connoisseur model for design assessment, see Richard Coyne and Anthony Snodgrass, *Interpretation in Architecture*, London and New York: Routledge, 2006, pp. 119–30.

20 Ideally, the process of generating quality statements should involve all those in the community of practice, including: new and experienced tutors, students and external examiners.

Chapter 6

Leslie Martin and the science of architectural form

Adam Sharr

I dedicate what follows to my former colleagues in architectural practice, who teased me when I turned to The Architect's Pocket Book *for reference and who also borrowed it when they thought I wasn't looking.*

This chapter is about a scientific model of architectural quality that was particularly influential in Britain during the late 1960s and early 1970s. It was promoted by institutional figures who sought to understand architectural design through the lens of science. They also sought to promote education over training for architects, consolidating the discipline of architecture in universities after winding down office-based pupillage.[1] Their research-led agenda had rational priorities – emphasising investigations of building types, functional needs, environment and climate, pre-fabrication systems, material properties and sociological and psychological analyses.[2] It was suggested that practices and universities should gather, analyse and deploy data of this kind, a sort now most familiar to readers of *Neufert's Architectural Data*, *The Metric Handbook* and *The Architect's Pocket Book*.[3] Theirs was a collegiate view of architectural endeavour – a long way from the myth of the heroic architect as creative genius – with teams of professionals seeking the logical, collaborative progress of the discipline in service of an orderly society.

Few devotees of this scientific model of architecture remain. Subsequent generations found it an easy target. A swathe of younger practitioners, among whom the Archigram group are now most famous, found it rigid and stifling.[4]

Phenomenologists criticised its inability to deal in sensory and emotional experiences, and its subjugation of the architect's individual creative judgement behind the operation of prescribed systems.[5] Critical theorists rallied against its elitism and essentialism, challenging its explicit hierarchies of professionals and inhabitants, and questioning both its tendency to treat individuals as 'users' and the singular authority that it accorded to functional 'truths'.[6]

In charting the shifting sands of architectural quality, however, this scientific model remains of distinctive importance. Its chief promoter in Britain – through his practice and the Centre for Land Use and Built Form Studies that he founded at the University of Cambridge – was Leslie Martin. In 1973, Martin was presented with the RIBA's Royal Gold Medal. Sponsoring his award, Trevor Dannatt argued that:

> As Peresutti said, the architect is 'the leader of a profound love of quality', and it is for this reason that we must respect Leslie Martin. And we must start with the quality of thought as underlying all.[7]

This paper explores Martin's thoughts – specifically the scientific model of architecture that he promoted – and examines his distinctive appreciation of quality.

Institutional influence

To understand the influence of Martin's model of architecture, it is important to appreciate his personal influence. Professor Sir Leslie Martin was perhaps the consummate institutional figure of British architecture in the 1960s and early 1970s.[8] Born in 1908, he qualified as an architect in 1931, winning on the way the RIBA's Silver Medal student prize and the Soane Medallion. After a spell as head of the Hull School of Architecture in the 1930s (a post to which he was appointed at the young age of 26), Martin spent the war years with the Architects' Department of the LMS Railway, pioneering prefabricated construction.[9] In the late 1940s, as Deputy Architect of the London County Council – at that time Britain's largest architectural practice – he jointly oversaw construction of the Royal Festival Hall, a building widely lauded as a symbol of a re-emboldened Britain's emergence from wartime privations.[10] Promoted to Architect of the LCC in 1953, Martin left after only three years to take up Cambridge University's first chair in architecture.[11]

At Cambridge, Martin lost no time in transforming the school and – in league with Richard Llewellyn Davies – reshaping British architectural education and formalising architectural research.[12] He masterminded and chaired the RIBA's influential Oxford Conference of 1958. Assuming the role of secretary to the event, Martin was able to record its support of his suggestion that architecture should become a graduate profession. He also recorded its endorsement of his view that postgraduate research into 'the advancement of architectural knowledge' should be formalised.[13]

Martin combined his Cambridge chair with an architectural practice that operated from his home, a converted mill near the university town, and sat on numerous government committees and competition juries. In the latter capacity, he helped to appoint Stirling and Gowan to the Leicester Engineering Building Commission and Jørn Utzon to Sydney Opera House. His completed projects included buildings for Oxford and Cambridge colleges, the Royal Scottish Academy of Music in Glasgow and the Gulbenkian Foundation in Lisbon. With Colin Buchannan, he also masterminded an unrealised plan for the wholesale rebuilding of Whitehall, London's government district, as a mat of stepped-section, high-density office slabs, dubbed 'ziggurats for bureaucrats' by Ian Rice.[14] Martin died in 1999. The sway that he exerted over the UK profession diminished in the 1970s, but the immense influence that he held for a time in professional, governmental and academic circles must not be underestimated. His role as a committee man gave him the opportunity to mould architectural research in Britain. It also afforded him opportunities to work towards the implementation of his scientific model of architectural practice among the profession.

Typology and reason

Martin sought to establish a body of theory that would be distinctively architectural, believing that university architecture departments should offer something to the intellectual commonwealth beyond contributions to the established fields of engineering and history. He was an advocate of architectural theory, but not as many would conceive of it today. To quote from his account of the RIBA Oxford Conference:

> The universities [. . .] have a right to expect that knowledge will be guided and developed by principles: that is, by theory. 'Theory,' as one speaker said, 'is the body of principles that explains and inter-relates all the facts of a subject.' Research is the tool by which theory is advanced. Without it, teaching can have no direction and thought no cutting edge.[15]

Theory, for Martin, was abstract and timeless. It was a matter of form, but not considered in relation to ornament, beauty or style. His preferred classification of form was in relation to building type, for example: housing, libraries or auditoria. To him, each type had its own logic. Every project for a housing scheme, or a library, had a site and a client, involving architectural judgement and a measure of site and client response. But behind those, for Martin, was an idealised housing-, or library-type form, which inhabited the realm of theory and informed any design proposition. Ideal type forms transcended style and architectural history, and were distinguished by inherent properties that integrated constructional, sociological and environmental factors.

In his 1983 monograph *Buildings and Ideas*, published by Cambridge University Press, Martin attributed his conception of type to nineteenth-century studies by Durand.[16] These studies – which were made by drawing to the same scale examples of theatres, stadia, market halls, and so on – demonstrated to Martin the characteristics of ideal forms, which he referred to as 'a system of axes, hierarchies and balances'. Martin conceived his types as rational. He measured their success in terms of functional efficiency, compact planning and economic construction. The terms order, coherence, reason and appropriateness were tokens of highest praise in his vocabulary. He wrote that:

> appropriate technical processes or constructional systems are rooted in and are inseparable from the needs of particular building types; second, that every building type [. . .] can be developed to the point at which its own organisation and therefore its characteristic forms can emerge [. . .] The characteristics of the type have first of all to be discovered.[17]

The perfection of type as ideal was an ultimate goal, and Martin's writings display his conviction that this ambition remained plausible. But his evocation of type was also, occasionally, tinged with the language of revelation and epiphany. Type forms were already there, ready to emerge and ripe for discovery. Due reverence paid to type was right and proper, for type was the ground of theory and the site of knowledge.

The quality of mathematical rigour

Two illustrations of Martin's theory of idealised typology come from his 1967 paper 'Architects' Approach to Architecture'. Concerned with density, economics and plot ratio, Martin and Lionel March carried out a series of abstract studies in association with the project for Whitehall's ziggurats.[18] Martin was impressed by Fresnel's diagram of squares (Figure 6.1) and claimed architectural insight from it. He wrote of this abstract diagram:

> Appearances can be deceptive. It is not easy to grasp at once, for example, that in the diagram of Fresnel squares, the outer annulus (in fact each annulus) has exactly the same area as the central square. And this is at the root of the problem. If these two forms were thought of as two contrasting ways to place a building on a site, it will be seen that questions of daylight, road access and the location of free space are widely different.[19]

Martin translated this into a series of generic forms exploring alternative ways to use a given site area. He tested permutations and combinations of tower,

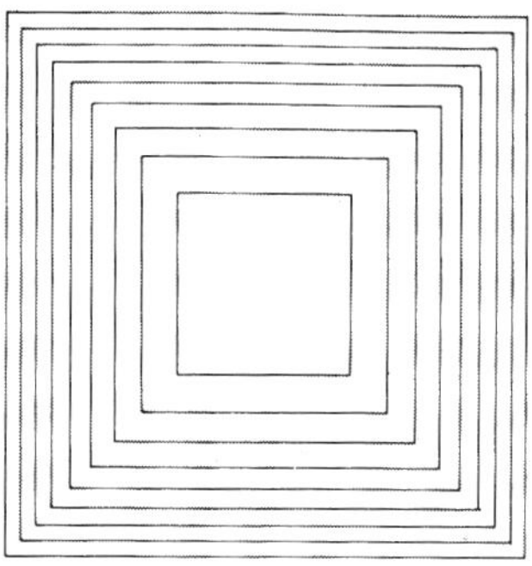

6.1
Fresnel's diagram of squares

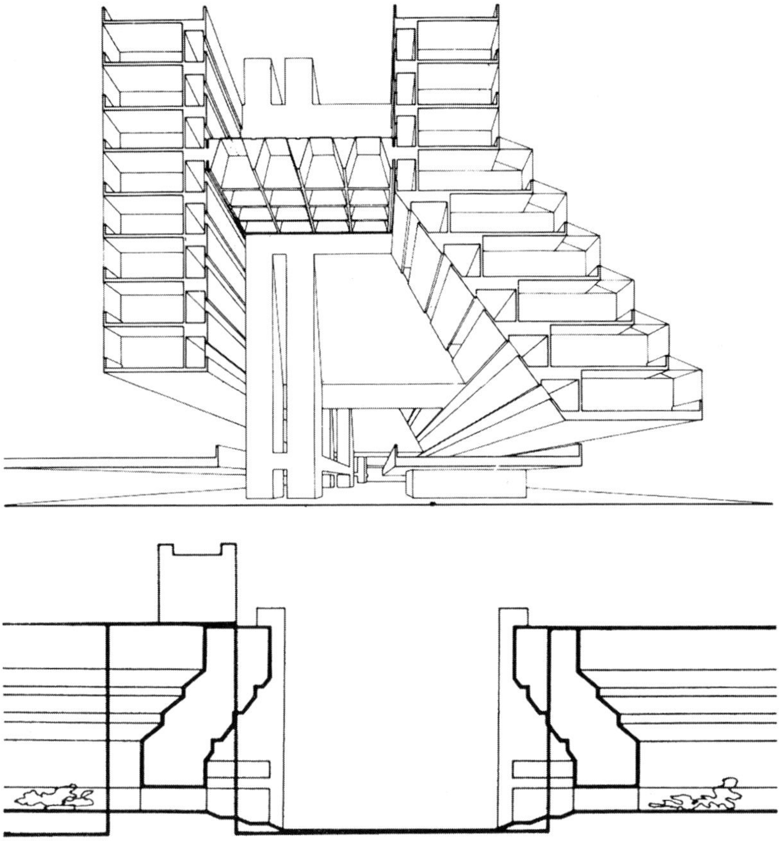

6.2
Sectional perspective and diagrammatic section of Leslie Martin's and Lionel March's project for Whitehall in London

slab and courtyard, examining their efficiency of land use in relation to form and public realm, and their basic environmental performance in terms of solar orientation, climate and internal environment. From Fresnel, Martin extrapolated a diagrammatic approach to the functional appropriateness of form. The plan and section forms of the Whitehall scheme (Figure 6.2) were consequently derived from a chosen archetype that he deemed to have appropriate density, economy and environmental performance.

In an era when towers and slabs set in open space were the default formal solution in Britain, Martin frequently advocated instead low-rise courtyard proposals, on the basis of matter-of-fact mathematical analysis. He felt that courtyard solutions frequently offered higher densities at lower cost – and maintained a more measured urban scale – but were ignored because of modernist architects' hot-headed ideological preferences. Martin's 1962 proposal, with Patrick Hodgkinson, for London's Foundling Estate in Marylebone (Figure 6.3) indicates how he generated land use principles from mathematical analysis:

> If the 15 per cent of office and commercial building were to be disposed around the perimeter of the site [. . .] it would occupy a band of buildings eight storeys high. It could turn a blind face to the road network: the distributor road could be placed inside it; the parking accommodated under it. Inside this again, the housing could be built in the form of one continuous five storey band or two discontinuous bands [. . .] that is at the scale of Georgian terraces. The remaining area of the site is now free for the sites of individual buildings. The hospital [. . .] can stand free in an open park-like surround. Within this inner park (about the size of St James' Park) schools and play spaces can be safely accommodated. An existing Georgian crescent, and even an area of Georgian housing, can be preserved if necessary.[20]

Martin's and Hodgkinson's Foundling was a logical world, orthogonally planned and derived from geometrical analysis. For Martin, such design solutions were distinguished by their mathematical rigour:

> When [the range of forms possible on a site] are supported by an elegant mathematical proof, that concept [of a spectrum of forms] is given tremendous power [. . .] the thought has been lifted from the particular to the level of principles which have general application.[21]

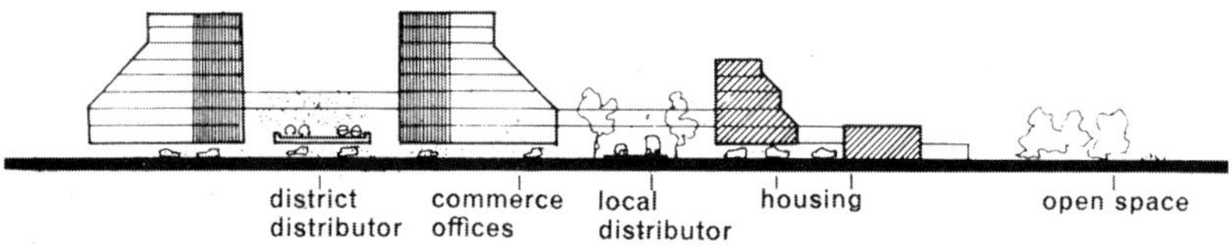

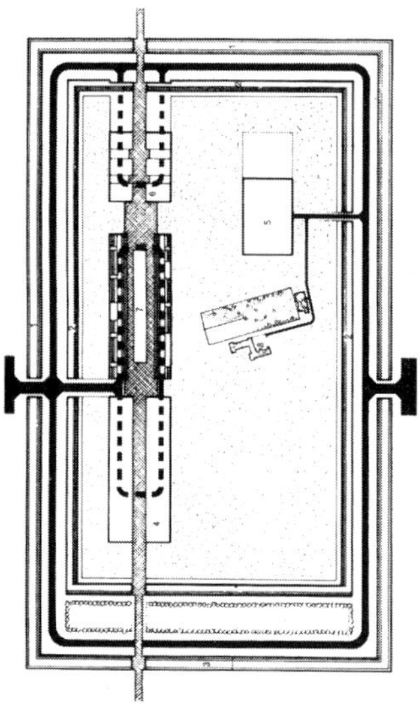

6.3
Section and plan of Leslie Martin's and Patrick Hodgkinson's project for the Foundling Estate in London

This model of architectural knowledge determined quality in relation to mathematical elegance. A good design proposition was attributed the coherence and intellectual brevity of a tightly formulated equation. Power and poise were to be found in reductive clarity.

It is important to understand Martin's model of architecture in its historical and intellectual contexts. The architects of the Foundling Estate made their proposals with the detached reserve of the cold war scientist. Technology at this time was not just the preserve of white coats, but also a source of pride and optimism in British culture at large – only two years later, Harold Wilson's famous 'white heat of technology' speech played a significant role in the 1964 election campaign. This was an era of seemingly unstoppable technological progress, which celebrated the conception of ever more liberating domestic devices while also dealing in ever more apocalyptic munitions.[22] Against this backdrop, Martin's can be read as a scientist's architecture, focused with deadly calm on the efficient production of space.[23]

Martin's designs, and his model of architecture, come as little surprise in the light of his Cambridge associations. The elite university remained, in the 1960s, a bastion of Anglo-American philosophy, pure and applied. Logical positivism remained for many the favoured *modus operandi*. Dean Hawkes reports, as one of the first graduate students in architecture there in the 1960s, that favoured sources in the department included Karl Popper's *Objective Knowledge*, P.B. Medawar's *The Art of the Soluble*, Christian Norberg-Schulz' *Intentions in Architecture* and Christopher Alexander's *Notes on the Synthesis of Form*.[24] Out of this *milieu*, under Martin's aegis, emerged Lionel March's studies in *The Geometry of Environment*, Philip Steadman's work on the morphology of building plans and on biological analogies for the evolution of designs, and Hawkes own work on *The Environmental Tradition*.[25] Martin allowed, and perhaps even encouraged, a measure of dissent, demonstrated by the presence at the outset of Colin St John Wilson and the young Colin Rowe, Peter Eisenmann and Catherine Cooke. But, philosophically, the subject knew its place in relation to the object here, in clear relation to a single, unrelativised truth. As any critical theorist would now point out, cultural differences were hardly of consequence in this world. There was little sense that technology might not be a decisive instrument of progress and salvation. At home in its time and place, Martin's project was no less than an emergent science of architectural form.

Feedback and dissemination

Unsurprisingly, Martin's projects are laid out according to type in his book *Buildings and Ideas*. He considered every building to be a research project, and the best buildings, to him, contributed to the development of the relevant type towards perfection. The knowledge of the architectural profession was thus constituted by best practice in each type, in relation to idealised land use, optimised functions, appropriate construction systems and balanced environmental performance. Each type, to him, should evolve in Darwinian fashion, the eccentricities and infelicities of particular building propositions testing and informing the efficacy of the generic.

Martin's model of architecture became influential in the British architectural profession and it was significant in the post-war reconfiguration of professional procedures. Its presence encouraged public architects' offices to think of each building task as a research project into type form. Until the early 1980s, post-war architectural practice in Britain was dominated by local authority offices; the greater proportion of architects worked for local and county councils, which had both time and inclination to invest in research into the building types on which they worked. From the LCC outwards, numerous local authority offices considered themselves – at least in part – to be research agencies.

The Martin approach, disseminated from Cambridge, was seen by many as a model to emulate. Its principles were supported, in large part, by the technically oriented Building Research Station, and they informed decisions taken by the Royal Institute of British Architects. The notion of post-occupancy evaluation – carried out by means of environmental monitoring, sociological questionnaires and the like – was written into the *RIBA Plan of Work*, the objective description of work stages according to which every job should proceed.[26] In this model, it would be a matter of routine that a building would be designed and built and its performance quantitatively analysed. Lessons would then be fed back into the knowledge base of the office ready for the next project. The parameters of this model were recorded in the UK's national adoption of a Swedish library system for architects, called CI/SfB,[27] and the preparation of a National Building Specification (NBS),[28] mechanisms still used by the British profession and that, along with the *Plan of Work*, are still defined by Martin's epistemology to a large extent.

This burgeoning scientific model of architectural endeavour was also adopted by the architectural media. Leslie Fairweather, editor of the *Architect's Journal* (*AJ*) at London's Architectural Press, was a key supporter. Where type form was advanced by an exemplary project, an objective evaluation was to be published for the collective benefit of architectural knowledge. Each issue of the *AJ* was structured around the provision of data (see Figure 6.4). A key component of this was a weekly building study disseminating knowledge on cost, engineering and environmental performance, as well as accounts of the project by client, architect, engineer and – last – a genuinely critical commentary by an expert on the building type. In this way, the *AJ* was to become the repository of the profession's current knowledge, its status among the profession equivalent to *The Lancet* for medical practitioners.

After the science of architectural form?

The *AJ* functioned in this way from the 1960s into the 1970s. But the scientific model of architecture that it reinforced – with its collegiate aspiration towards the development of professional knowledge – could not outlast the wholescale shift of practice from public to private sectors in Britain's Thatcherite 1980s and the ensuing commodification of expertise.[29] A private practice's knowledge became its selling point, to be guarded for its commercial value rather than disseminated. Information sharing ran counter to the practice of selling services, and post-occupancy evaluation – quietly removed from the RIBA's *Plan of Work* in the 1990s (although

6.4
1973 diagram showing typical contents of the *Architects' Journal*, drawn as an exploded isometric

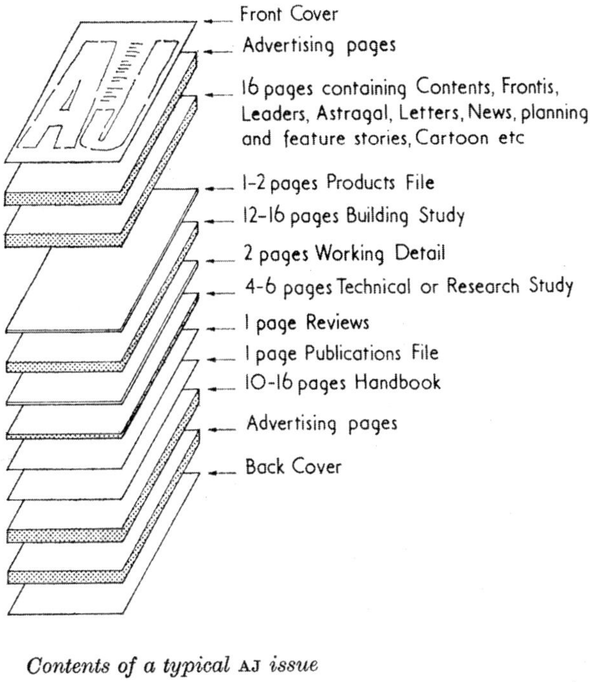

since reinstated) – became an all-too-public and potentially costly admission of professional inadequacies. This commercial shift paralleled shifts in academic priorities – with the intellectual ground of architecture schools moving towards postmodernism, phenomenology and critical theory – and Martin's science of architectural form drifted away from the centre of professional and academic consciousness. The limits of its positivist outlook, its tendency to abdicate political responsibility and its elitist focus on the professional have been made apparent. Martin, too, was perhaps aware of its limitations. A parallel interest in Aalto – who he nominated for the RIBA Gold Medal – seems to have informed a handful of his design projects (such as the Kettle's Yard gallery and Harvey Court at Cambridge), where a measure of controlled playfulness is occasionally allowed to subvert the rational.

Despite the decline in its influence, the legacies of Martin's science of architectural form remain pervasive. The construction industry – in the UK and beyond – prefers the positivist agenda that Martin played his part in institutionalising.[30] Economic rationality, optimum functional performance and quantifiable decision-making remain favoured modes of procedure. While it is often little understood and rarely articulated, the philosophical position underpinning industry structures remains, to a considerable degree, that advanced by Martin and his associates. Architectural delight, dangerously subjective, remains for the building industry a slippery quantity and a lesser priority. In construction culture, beyond the narrow frame of architects' discussions with architects, the largely unarticulated residues of Martin's epistemology remain decisive.

Within the frame of architectural debate, too, his model of knowledge based on functional study of type form survives. Arguably, it persists as a parallel discourse: one in which – it seems to me – architects tend to admit their participation only reluctantly. Designers now prefer to celebrate individual judgement over system. The cult of the individual tends to prevail in the journalist-trained architectural media and in public relations-savvy practice.[31] Priorities of education and publishing tend to support an emphasis on the individual. So too does the commercial imperative of selling expertise. Martin's model of practice, focused on calm evaluation and shared knowledge, fits badly with a professional opportunism that encourages the marketability of genius. But architects still make recourse to building type precedent, to abstract analysis of form and associated environmental studies, more than many of them would readily admit. In *Neufert's Architectural Data*, in *The Metric Handbook*, in *The Architect's Pocket Book* – regular sources of covert reference for practitioners – are the manifestations of Martin's influence. His influence is also acknowledged more openly in a few strands of contemporary design practice, notably in the re-invention of housing types by Dutch architects such as MVRDV.[32]

I would argue that the current professional judgement of Martin's model of quality finds it of ambiguous benefit. Information derived from functional studies of building types is considered helpful if treated as guidance rather than as some sort of truth. However, the idea of a singular objective-regulating system is found to stifle design creativity. To me, obsessions with perfected function ignore the functional redundancy that is often the grit in the oyster making for deliciously unquantifiable delight. However, the egalitarian ideal of sharing information – the idea of a profession engaged together in designing for a greater good – remains a worthy aim, even if its utopian goal must inevitably be out of reach. That said, the influence of a small, institutional elite working to emphasise a single model of practice – however egalitarian the intended aims of that model – should be resisted.

For Martin, quality was not a designation but an inherent property of form, many of whose parameters could be quantified through geometrical properties or mathematical and sociological data. Those architectural qualities that are felt rather than measured are often of lesser consequence here. Meaning is almost meaningless. It has been fashionable among certain architects to despair about this positivist ethos, disparaging it among themselves while persisting with it in order to talk the commercial talk with professional clients, project managers and quantity surveyors. What few architects in this position remember is that – courtesy of Leslie Martin and his associates – this pervasive ethos is, at least in part, of architects' own making.

Notes

1 Leslie Martin, 'Conference on Architectural Education', *RIBA Journal*, 1958, vol. 6, pp. 279–82; Mark Crinson and Jules Lubbock, *Architecture: Art or Profession? Three Hundred Years of Architectural Education in Britain*, Manchester: Manchester University Press, 1994.

2 Martin appears to have been less interested in the psychological aspects of research at this time than aspects of function and climate. On the development of architectural psychology in the 1960s, see: Alise Upitis, 'Personality, Consumption and Architectural Pedagogy in the UK, 1958–1968', *arq*, 2008, vol. 12, pp. 3–4.

3 Ernst Neufert, *Architectural Data*, Oxford: Blackwell Science, 1995; David Littlefield (ed.), *Metric Handbook: Planning and Design Data*, Amsterdam: Elsevier/Architectural Press, 2008; Charlotte Baden-Powell, Jonathan Hetreed and Anne Ross, *Architect's Pocket Book*, Oxford: Architectural Press, 2008.

4 Simon Sadler, *Archigram: Architecture without Architecture*, Cambridge, MA: MIT Press, 2005.

5 Juhani Pallasmaa, *Eyes of the Skin: Architecture and the Senses*, London: Academy Editions, 1996.

6 K. Michael Hays, *Architecture Theory Since 1968*, Cambridge, MA: MIT Press, 1999.

7 Trevor Dannatt, 'Trevor Dannatt Praises the Profound Love of Quality', *RIBA Journal*, August 1973, pp. 382–3 (p. 382).

8 Peter Carolin and Trevor Dannatt (eds), *Architecture, Education and Research: The Work of Leslie Martin – Papers and Selected Articles*, London: Academy Editions, 1996.

9 Trevor Dannatt, 'Early Years: Towards a New Order', in Carolin and Dannatt (eds), *Architecture, Education and Research*, pp. 12–25.

10 John McKean, *Royal Festival Hall: London County Council, Leslie Martin and Peter Moro*, London: Phaidon, 1992.

11 Dean Hawkes, 'The Cambridge School:Tradition, Research, Development and the Martin Centre', in Carolin and Dannatt (eds), *Architecture, Education and Research*, pp. 122–7.

12 Martin, 'Conference on Architectural Education'; Leslie Martin, 'Education Around Architecture: Address to the RIBA Conference, Birmingham, 1970', *RIBA Journal*, September 1970.

13 Martin, 'Conference on Architectural Education'.

14 Ian Rice, 'Ziggurats for Bureaucrats': Sir Leslie Martin's Whitehall Plan, *arq*, 2004, vol. 8, pp. 3–4, pp. 313–23; Leslie Martin and Colin Buchannan, *Whitehall: A Plan for the National and Government Centre*, London: HMSO, 1965.

15 Martin, 'Conference on Architectural Education'.

16 Leslie Martin, *Buildings and Ideas 1933–83: From the Studio of Leslie Martin*, Cambridge: Cambridge University Press, 1983, p. 15; J.N.L. Durand, *Recueil et parallèle des édifices*, Paris: n.p., 1801.

17 Martin, *Buildings and Ideas*, p. 14.

18 Martin, *Buildings and Ideas*, pp. 120–7.

19 Martin, 'Architects' Approach to Architecture'.

20 Martin, 'Architects' Approach to Architecture'.

21 Leslie Martin, 'Leslie Martin on the Bridges Between the Cultures', *RIBA Journal*, August 1973, pp. 381–8 (p. 383).

22 Matthew Hilton, 'Consumer Politics in Post-War Britain', in Martin Daunton and Matthew Hilton (eds), *The Politics of Consumption: Material Culture and Citizenship in Europe and America*, Oxford: Berg, 2001.

23 Mark P. Donnelly, *Sixties Britain: Culture, Society and Politics*, Harlow: Longman, 2005.

24 Karl Popper, *Objective Knowledge: An Evolutionary Approach*, Oxford: Clarendon Press, 1972; P.B. Medawar, *The Art of the Soluble: Creativity and Originality in Science*, Harmondsworth: Penguin, 1969; Christian Norberg-Schulz, *Intentions in Architecture*, London: Allen & Unwin, 1963; Christopher Alexander, *Notes on the Synthesis of Form*, Cambridge, MA: Harvard University Press, 1964. Thomas Kuhn's *The Structure of Scientific Revolutions* also became influential in Martin's circle, although its publication coincided with the professor's retirement in 1973. I am grateful to Dean Hawkes for this information, given in conversation, which he has also written about: Dean Hawkes, 'Sustainability and Architectural Synthesis', in *Architecture City Environment: Proceedings of PLEA 2000*, Cambridge, UK, 2–5 July 2000, London: James & James, 2000.

25 Lionel March and Philip Steadman, *The Geometry of Environment: An Introduction to Spatial Organisation in Design*, London: RIBA, 1971; Philip Steadman, *Evolution of Designs: Biological Analogy in Architecture and the Applied Arts*, Cambridge: Cambridge University Press, 1979; Philip Steadman, *Architecural Morphology: An Introduction to the Morphology of Building Plans*, London: Pion, 1983; Dean Hawkes, *The Environmental Tradition: Studies in the Architecture of Environment*, London: E&FN Spon, 1995.

26 *Royal Institute of British Architects Plan of Work*, 1st edition, London: RIBA, 1967.

27 'Guide to the *AJ* Information Library', *AJ*, 11 October 1972, vol. 41, p. 156, pp. 833–48.

28 *National Building Specification 1973*, 1st edition, London: NBS Ltd, 1972.
29 David Marsh and R.A.W. Rhodes, *Implementing Thatcherite Policies: Audit of an Era*, Buckingham and Philadelphia, PA: Open University Press, 1992.
30 C.G. Powell, *The British Building Industry Since 1800: An Economic History*, London: E&FN Spon, 1996.
31 Anna Klingmann, *Brandscapes: Architecture in the Experience Economy*, Cambridge, MA: MIT Press, 2007.
32 *MVRDV Files: Projects 002–209*, Tokyo: A+U Publishing, 2002.

Chapter 7

Ugly little angels revisited

Matthew A. Cohen

The nave arcades of the Basilica of San Lorenzo, those rhythmic, monumental assemblages of columns, entablature blocks and arches, in the early fifteenth century established a benchmark of elegance and refinement for the new Renaissance style (Figure 7.1). As complete compositions, the nave arcades suppress the individuality of their component parts. From afar, the capitals and other carved details all appear identical, but, from close up, they betray notable variations in quality of design and execution. These variations reflect the compelling circumstances under which this pivotal work in the history of architecture came into being.[1]

Variable quality

The sculptural embellishments of the nave arcades lend to the basilica much of its opulent character. Opulence, however, is more evident in some areas than others. In the three western bays of the nave, the sculptural carvings – which depict leaves, flowers, lambs and winged angels – are executed with a high degree of naturalism and refinement (Figures 7.2 and 7.4). In the five eastern bays, by contrast, these elements appear, not only different in style, but significantly less refined – occasionally even crude – by comparison (Figures 7.3 and 7.5). My measurements of the arcades, furthermore, reveal extremely precise dimensional consistency among the three western bays, but substantial irregularity in the five eastern bays.[2]

Physical evidence suggests that the builders recognized these lapses in quality and tried to mitigate them. In the five eastern bays, for example, the most egregious lapses in quality in the ornamental carvings face the dark, northern side of the north arcade, where they can never clearly be seen. In those bays, furthermore, some components contain repairs (such as a new top on an otherwise

7.1

Basilica of San Lorenzo north nave arcade, view looking east

7.2
San Lorenzo nave arcades phase I, column 6

7.3
San Lorenzo nave arcades phase II, column 11

7.4
San Lorenzo nave arcades phase I, column 9, frieze block detail

7.5
San Lorenzo nave arcades phase II, column 11, frieze block detail

monolithic column shaft) or dimensional compensations (such as an unusually short entablature block atop an unusually tall column) that are not found in the nearly perfectly executed three western bays.[3]

Documentary evidence indicates that the nave arcades were built in two distinct phases. The first, encompassing the three western bays, lasted from 1446 to 1450. The second, encompassing the five eastern bays, probably began around 1463, but the dates are otherwise uncertain.[4] Volker Herzner and Gabriele Morolli believe that the second phase was not begun until after 1465, under the patronage of Cosimo's son, Piero de' Medici. Morolli furthermore attributes the differences in style between the two phases to the new aesthetic preferences of the skilled craftsmen who arrived with the change in patronage. He attributes what he perceives to be a progressive decline in sculptural quality during the course of the second phase to a 'growing disinterest' on the part of Piero, and later Piero's son Lorenzo, in the early quattrocento style of the basilica.[5]

In a previous study, conversely, I proposed that the second phase was entirely completed by October 1464, and thus under Cosimo's patronage, like the first phase. In that study I attributed the differences in style and quality between the two phases to a conscious decision on the part of Cosimo to sacrifice quality for speed and rush the basilica to completion.[6] After additional research and reflection, I now propose the revised hypothesis that Cosimo probably initiated the second phase of nave construction in 1463, but did not complete it before he died the following year. While available evidence indeed suggests that Cosimo's contribution to this phase included some of the redesigned, lower-quality ornaments, this interesting possibility remains uncertain. This revised hypothesis is not inconsistent with that of Morolli, for, regardless of who initiated the new manner of sculptural carving in the five eastern bays of the nave, each patron may have had his own particular motives for bringing the work into the form we see today. The present study explores the historical circumstances that accompanied and perhaps helped to shape the remarkable change in quality between the two phases of nave construction in the Basilica of San Lorenzo.

Cosimo's ascendancy

Construction of the basilica began in the transept in 1421, following the design and supervision of the architect-prior Matteo Dolfini. The project was funded by the church, prominent citizens and the *comune*. After about a year, probably at the death of Dolfini, Giovanni de' Medici took control of the project, assuming much of the expense and installing Filippo Brunelleschi as the new *capomaestro*. Brunelleschi redesigned the basilica on Dolfini's foundations. Construction largely stopped in about 1425 owing to wartime economic pressures, but the Old Sacristy and adjacent double chapel (Figure 7.6) continued to completion under secure Medici funding in 1428.[7]

In November 1440, noting that the incomplete and deteriorating new transept was 'a source of humiliation and shame for the entire population of the parish', church officials offered the prestigious rights of patronage of the high altar chapel to whomever would agree to pay for its construction.[8] It was a thinly veiled

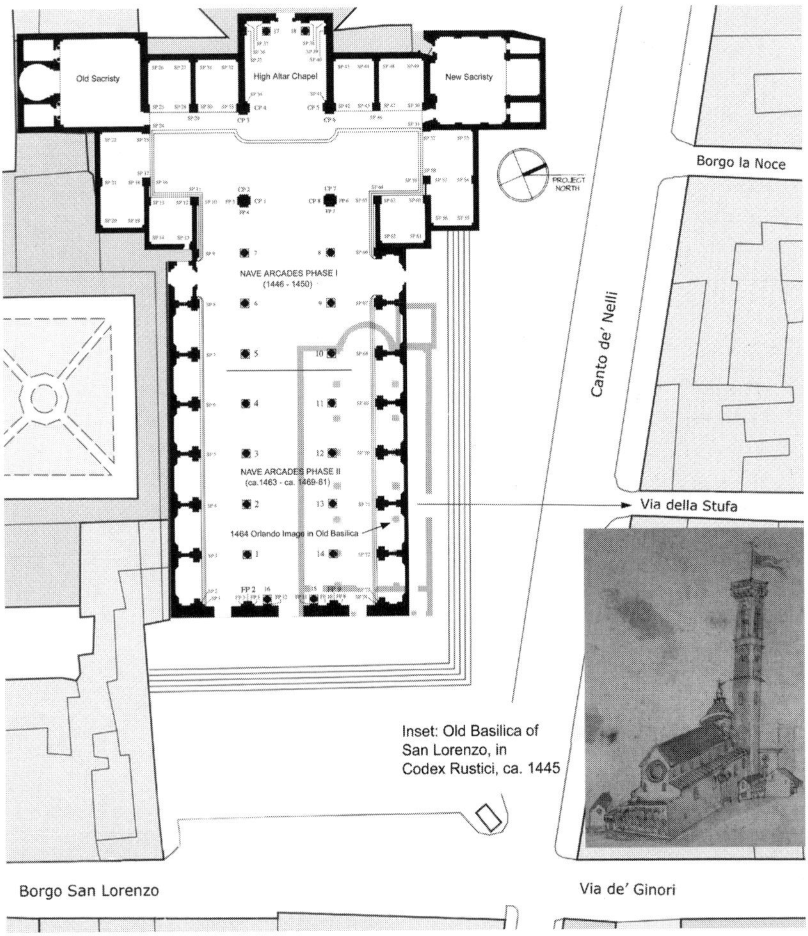

7.6
Basilica of San Lorenzo, floor plan with column and pilaster numbers, nave construction phases and reconstructed footprint of the old Basilica of San Lorenzo, in existing conditions site plan.
Inset: Old Basilica of San Lorenzo shown in Codex Rustici, c. 1445 (Seminario Archivescovile, Florence)

offer to the only person capable of accepting it – Giovanni's son, Cosimo. Perhaps both to fulfill his father's commitment and to elevate his own status, Cosimo accepted the offer, and then some. In a notarial record of August 1442, the church conceded to him, not only the rights of patronage to the high altar chapel, but also to 'the nave in the middle of the church, extending as far [east] as the high altar of the old church'. The offer was made on the condition that Cosimo complete the work within six years.[9] Since extensive evidence locates the old Romanesque basilica and cloister of San Lorenzo on the site of the five easternmost nave bays (Figure 7.6), and since a Medici construction ledger records payments between 1446 and 1450 for 'six large columns of *macigno*' and other components of the three westernmost bays of both nave arcades, there can be little doubt that these three nave bays were completed by 1450.[10]

Cosimo's anxiety

Cosimo's deadline for completion of the agreed-upon portions of the basilica, August 1448, appears to have passed without consequence, and the final piece of his commitment, the crossing dome, was not closed until 1456.[11] Whatever the cause of the delay – perhaps simply an unrealistic construction estimate made back in 1442 – Cosimo, judging from the consistently high quality of work found throughout the basilica west of columns 4 and 11, does not appear to have been particularly concerned about time. That would soon change.

At the completion of the crossing dome, Cosimo was sixty-seven years old. He must have been well aware that two other prominent men of his time, his father, Giovanni, and Brunelleschi, both lived to just sixty-nine. Although his formal obligations to the church had now been fulfilled, for he had completed the new church as far east as the high altar of the old church, Cosimo pressed on with his patronage. One year after the church canons held a supper in May 1456 for the workers who were about to close the crossing dome, they held another to celebrate the 'beginning of the construction of the new cloister'.[12] Why would Cosimo, who must have been anxious to secure the completion of this basilica – his final resting place – within his lifetime, now proceed to the construction of the canons' residences rather than the remaining portions of the basilica?[13] According to the fifteenth-century biographer Vespasiano da Bisticci, when Cosimo was asked this very question, he replied that, if he did not build the residences, no one would, 'because there would be many who would want to build the church, but not the cloister, there being [in it] much greater prestige'.[14] Evidently, Cosimo doubted whether he would live long enough to see the completion of his vision for a new San Lorenzo complex, and whether his family would be able to complete the work after his passing.

Cosimo had begun to contemplate his legacy, and his outlook was gloomy. 'I know that after my death . . .' he once said, according to Vespasiano, '. . . my children will be in worse condition than those of any other Florentine who has died for many years past'.[15] Cosimo had good reason to worry for the future of his family's continued wealth and prominence. He must have foreseen the impending decline of the Medici bank, for, by the time of his death on 1 August 1464, de Roover writes, 'his company had passed the peak of its prosperity and was going downhill. The London branch had come to grief; other branches, too, were experiencing growing difficulties; and profits were falling off'.[16] Cosimo would be succeeded as head of the Medici bank by his son Piero, who, despite his lack of training in finance, would see fit to order a survey of the family assets in order to see 'in how many feet of water he was standing'.[17] In addition to these uncertain business prospects, during the last years of Cosimo's life, Piero's succession to the position of de facto head of state was in doubt, and, indeed, the two men hardly projected an image of dynastic security. Both were bedridden much of the time, leading one of Cosimo's former supporters to call them 'cold fish [*huomini freddi*] whom illness and old age have reduced to such cowardice that they avoid anything that might cause them trouble or worry'.[18] Medici foes lay in wait, with Medici power apparently resting solely on the shoulders of the ailing Cosimo. When asked to join in a conspiracy against Cosimo, Palla Strozzi noted in 1460 that such a scheme would be ill-advised, for, according to Rubinstein, 'as long as Cosimo was alive, it would be impossible to get rid of him; [but] once he was dead, within a few days conditions would develop according to [his enemies'] wishes'.[19] Cosimo must have heard these rumblings.

Vespasiano claims to have once heard Cosimo lament that

> one of the greatest mistakes of his life was that he did not begin to spend his wealth ten years earlier, because, knowing well the nature of his city, he was sure that within not even fifty years, no memory of himself or of his house would endure save the few monuments he might have built.[20]

Might Cosimo have been thinking about San Lorenzo, which, by the time the crossing dome was completed, was running eight years behind schedule? Vespasiano tells us that, concurrent with the church and cloister of San Lorenzo, Cosimo completed the cloister and much of the church of the Badia of Fiesole, noting: 'He pushed this edifice to completion with all possible haste, always doubting that his time would be sufficient.'[21] Later, according to this account, Cosimo contracted with Vespasiano himself, a noted Florentine bookseller, to create an extensive library at the Badia by arranging for the copying of manuscripts. Here too, writes Vespasiano, 'his wish was that it be completed with all possible haste, and money was no object'.[22]

Cosimo's delay

In August 1461 the cloister was finished, the high altar of the new basilica was consecrated, and, in a solemn procession, the sacred relics of four saints (including Saint Lawrence) were transferred to it from the old basilica.[23] Cosimo was now seventy-two and had probably already lived longer than he had expected. The transept had taken about six years to complete; the dome, about five years; the first three bays of the nave, four years; and the cloister, just over four years. At this rate, Cosimo could expect the remaining work to require another four to five years, but he perhaps knew that he would not live that long. With money apparently not yet lacking (in light of Vespasiano's preceding comment), Cosimo would seem to have had ample personal motive to proceed immediately with construction of the remainder of the nave, at an expedited pace. Indeed, the canonry and entire parish would seem to have had ample motive as well. In August 1461, the newly consecrated basilica would have had a gaping hole in the end of the incomplete nave, extending from roof to floor, immediately east of columns 5 and 10 (Figure 7.6).[24]

Demolition of the old basilica and completion of the new nave would thus seem to have been the next most logical and expedient course of action. Indeed, Vespasiano seems to report just such an eventuality when he notes: 'Having finished the cloister [Cosimo] commenced the continuation of the church, and finished a good portion of it before he died.'[25] Since the only part of the basilica remaining to be built after the completion of the cloister in 1461 was the nave east of columns 5 and 10 (Figure 7.6), and since the old basilica stood in the way, we might logically conclude that the old basilica was demolished soon after the consecration of the new high altar, also in 1461. A seemingly contradictory piece of evidence, however, complicates this conclusion.

The testament of Orlando di Giovanni d'Orlandini, dated 9 October 1464, provides for the maintenance of a lamp in the 'church of San Lorenzo of Florence' to illuminate an image of the Virgin Mary,

> situated on the second column to the right of the entrance to the said church, or at the column, which is in said church, closer by and near the door through which one goes out and proceeds in a straight line to Via della Stufa.[26]

The present basilica does not have a door that opens 'in a straight line' to Via della Stufa, but the old basilica did, according to both a detailed fifteenth-century view and my reconstruction of the old basilica floor plan on the site (Figure 7.6).[27] There would seem to be but one way to reconcile both Vespasiano's indication that Cosimo continued construction of the new basilica soon after August 1461 and the continued existence of the old basilica in October 1464: Cosimo must have proceeded only with the remaining portion of the southern half of the nave.

The old basilica appears to have been a small, narrow structure that would not have significantly impeded construction of the remaining southern nave chapels and arcade bays (Figure 7.6).[28] Finishing the southern part of the nave before the northern part would have permitted Cosimo to leave the old basilica, and its numerous private chapels, undisturbed until new space into which many of those chapels could be transferred was ready. Proceeding as such may have solved another problem as well.

Cosimo appears to have been free to start construction in the southern part of the nave as soon as the new cloister was finished in August 1461. Only on 6 April 1463, however, does a church ledger record the beginning of payments by five families for construction of the southern nave chapels as a joint project 'on behalf of and for the pleasure of the venerable Cosimo de' Medici'.[29] Completion of the southern half of the nave could not have proceeded without these chapels. They could not have been added later, since the new cloister blocked access from the south, and they appear to serve as structurally integral buttresses to the side aisle vaults.[30] The reason for the twenty-month delay may be that Cosimo, like his father before him, had difficulty assembling enough chapel patrons.[31] An indication of this difficulty is provided by a church document of April 1465 that concedes to Piero de' Medici the authority to assign patronage of the unbuilt northern chapels to whomever he pleased, in order to expedite their completion.[32] By proceeding with the remaining southern part of the nave before the northern part, Cosimo only had to find five more chapel patrons, rather than ten.

In the sixteen months from 6 April 1463, when construction of the remaining southern nave chapels began, to Cosimo's death on 1 August 1464, we may reasonably expect that at least some of the pilasters that frame the chapel openings (along with their lower-quality capitals) took shape. Since Cosimo appears to have been anxious to complete the basilica as quickly as possible, we might also reasonably expect that the corresponding portions of the southern nave arcade (along with their lower-quality capitals and entablature blocks) took shape simultaneously, within Cosimo's lifetime and under his patronage. Definitive evidence to prove such a chronology, however, is lacking. I propose, therefore, as a modification to Morolli's hypothesis, that Cosimo rather than Piero might have been the first to approve the lower-quality sculptural embellishment. By either scenario, after 1461, both men had ample reason to value speed over quality: Cosimo wanted to complete the basilica before he died, and Piero simply wanted this costly, inherited obligation finished once and for all.

Available documents pertaining to the construction of some of the northern nave chapels date from 1469 to 1479, but physical evidence suggests that the nave was completed well before the last of them.[33] The angels carved into the entablature blocks of columns 11 and 12 appear to be the products of the same *bottega* and to depict the same children's likenesses, as do those of the entablature blocks of floor pilasters 2 and 9 on the interior façade. Thus, all four entablature blocks appear to have been made at about the same time, possibly many years prior to the completion of the northern nave chapels.[34]

Time, cost and quality

The construction chronology proposed here, which has Cosimo serving as the patron for both the high-quality sculptural embellishment of the first phase of nave construction and perhaps some of the lower-quality embellishment of the second phase, implies that Cosimo possessed a substantial degree of aesthetic open-mindedness, apparently motivated by pragmatism. This proposed chronology now raises the additional question of whether the master mason who played a leading role in the construction of the first phase of nave construction shared Cosimo's open-mindedness. Pagno di Lapo Portigiani, who probably oversaw day-to-day operations after construction recommenced in 1442, is noted in a document of July 1462 as '*capomaestro* at San Lorenzo'.[35] If he remained in that position nine months later, when construction of the remaining southern nave chapels commenced, then we would seem to have the remarkable situation in which both the same patron and same construction supervisor oversaw a dramatic reduction in construction quality within one building, evidently in order to accelerate its completion.

How far this basilica had fallen: from Brunelleschi's revolutionary design and sculptural finesse that drew marveling onlookers to the construction site, to a symbol of the wealth and influence of an increasingly confident Cosimo, to a record of a declining House of Medici, including Cosimo worrying over the future of his family and the very security of his tomb, and Piero struggling to complete the basilica while keeping the family business afloat.[36] This human drama is written into the stones of the basilica, alongside the more deliberately conceived iconography of form and ornament. While Cosimo's primary concern appears to have been not money, but time, Piero's may well have been equally money and time. Both men, however, appear to have evaluated time or money, or both, against quality, and to have chosen to sacrifice quality.

Notes

1 This essay is an abridgement and partial revision of my paper: 'Ugly Little Angels: Deliberately Uneven Construction Quality in the Basilica of San Lorenzo in Florence', *arq*, Dec. 2007, vol. 11, no. 3/4, pp. 277–89. I thank the Soprintendenza per i Beni Architettonici e per il Paesaggio per le Provincie di Firenze, Pistoia e Prato, for allowing me access to the basilica of San Lorenzo and the use of a lightweight mobile scaffolding from November 1991 to June 1992; Prior Mons. Angelo Livi and the chapter of San Lorenzo for their collaboration; James S. Ackerman, Howard Burns and Pietro Ruschi for advice and encouragement; Jack Wasserman for his incisive comments on the manuscript; and the Graham Foundation for Advanced Studies in the Fine Arts for its generous support.

2 In the three western bays, to take one example of dimensional consistency, column heights vary by no more than 1.6 cm from one to the next. In the five eastern bays, these heights vary by as much as 14 cm. Matthew A. Cohen, 'How Much Brunelleschi? A Late Medieval Proportional System in the Basilica of San Lorenzo in Florence', *Journal of the Society of Architectural Historians*, March 2008, vol. 67, no. 1, p. 47 and Appendices 2 and 4; and Cohen, 'Ugly Little Angels', 2007, p. 279, Fig. 4.

3 For more detailed analysis of the differences between these two portions of the nave, see Cohen, 'Ugly Little Angels', 2007, pp. 277–83 and 286; Gabriele Morolli, 'L'ordine brunelleschiano: morfologia e proporzioni', in Gabriele Morolli and Pietro Ruschi (eds), *San Lorenzo, 393–1993, l'architettura: le vicende della fabbrica*, Florence: Alinea, 1993, pp. 81–94; and Morolli, 'Non solo Brunelleschi: San Lorenzo nel Quattrocento', in Timothy Verdon (ed.), *Alla riscoperta delle chiese di Firenze: 5. San Lorenzo*, Florence: Centro di, 2007, pp. 94–107.

4 For documentation supporting these dates, see notes 9 and 28.

5 Volker Herzner, 'Zur Baugeschichte von San Lorenzo in Florenz', *Zeitschrift fur Kunstgeschichte*, 1974, vol. 37, pp. 97–108; and Morolli 'Non solo Brunelleschi', 2007, p. 102.

6 Cohen, 'Ugly Little Angels', 2007, p. 286.

7 A document of 1440 notes that construction of the basilica had been at a standstill for 'about fifteen years'. Piero Ginori Conti, *La Basilica di S. Lorenzo di Firenze e la famiglia Ginori*, Florence: Fondazione Ginori Conti, 1940, p. 237. For other sources on the early construction history, see Cohen, 'How Much Brunelleschi?', pp. 41–3.

8 Ginori Conti, *La Basilica di S. Lorenzo . . .*, 1940, pp. 236–40.

9 Ibid., pp. 240–5. The notarial record must have been a formality, for Cosimo began funding for quarrying work for the basilica on March 24, 1442 (modern style). Isabelle Hyman, 'Fifteenth Century Florentine Studies: The Palazzo Medici; and a Ledger for the Church of San Lorenzo', Ph.D. dissertation, New York University, 1968 (Ann Arbor, MI, 1970), pp. 305 and 431; and Hyman, 'Notes and Speculations on S. Lorenzo, Palazzo Medici, and an Urban Project by Brunelleschi', *JSAH*, May 1975, vol. 34, no. 2, p. 98.

10 On the footprint of the old basilica, see discussion below, and Cohen, 'How Much Brunelleschi?', Appendix 8. On the construction dates of 1446–50 for the first phase of nave construction, see Isabelle Hyman, 'Fifteenth Century Florentine Studies . . .', Ph.D., pp. 337, 341–5, 349, 352, 358–61, 475, 484, 496, 500, 511, 514, 516, 538–43, 546; and Hyman, 'Notes and Speculations', 1975, p. 117, Document 41.

11 Saalman, *Filippo Brunelleschi: The Buildings*, University Park: Pennsylvania State University, 1993, p. 439, Document 9.

12 Domenico Moreni, *Continuazione delle memorie istoriche dell'ambrosiana imperial basilica di S. Lorenzo di Firenze*, 2 Vols., Florence, 1816, Vol. 1, p. 14.

13 On the Basilica of San Lorenzo as Cosimo's mausoleum, see Irving Lavin, 'Donatello's Bronze Pulpits in San Lorenzo and the Early Christian Revival', in *Past–Present*, Berkeley, CA: University of California Press, 1993, pp. 1–27; and Dale Kent, *Cosimo de' Medici and the Florentine Renaissance*, New Haven, CT and London: Yale University Press, 2000, pp. 377–84.

14 The complete passage is as follows:

> *Domandato perché egli cominciava prima la casa che la chiesa, rispondeva che non sarebbe chi lo facessi, perché in molti sarebbono che farebono fare la chiesa, che non farebono fare la casa, sendo di magiore riputatione. Finita la casa, comincio a seguitare la chiesa, e fenne una buona parte inanzi che morissi.*

> Asked why he began first the cloister rather than the church, he replied that there would be no one who would do it, because there would be many who would want to build the church, but not the cloister, there being [in it] much greater prestige. Having finished the cloister, he commenced the continuation of the church, and he finished a good portion of it before he died.

Aulo Greco (ed.), *Vespasiano da Bisticci, Le vite*, 2 Vols., Florence: Istituto nazionale di studi sul rinascimento, 1976, Vol. 2, p. 182. I do not interpret this passage, as does Elam, as an indication that Vespasiano believed that Cosimo began the cloister as 'his initial act', before building any other part of the basilica complex; Caroline Elam, 'Cosimo de' Medici and San Lorenzo', in Francis Ames-Lewis (ed.), *Cosimo 'il Vecchio' de' Medici, 1389–1464: Essays in Commemoration of the 600th Anniversary of Cosimo de' Medici's Birth*, Oxford: Clarendon Press, 1992, p. 174. In the fifteenth-century context in which this comment was recorded, it would have been sufficiently obvious to everyone in Florence that, since 1442, Cosimo had undertaken construction of the basilica as a most extraordinary act of private patronage (see note 8). What may have struck many as puzzling, however, is why he would have temporarily left the impressive new basilica incomplete in order to build the cloister. I would furthermore note that Vespasiano uses the word *seguitare* to describe Cosimo's work on the basilica after he completed the cloister, a word that usually implies continuation.

15 Vespasiano da Bisticci, ibid., p. 192.

16 Raymond de Roover, *The Rise and Decline of the Medici Bank: 1397–1494*, New York: W.W. Norton, 1966, p. 358.

17 Ibid., pp. 358–9.

18 Nicolai Rubinstein, *The Government of Florence Under the Medici (1434 to 1494)*, 2nd edition, Oxford: Oxford University Press, 1997, p. 155; and Alison Brown, 'Piero's Infirmity and Political Power', in Andreas Beyer and Bruce Boucher (eds), *Piero de' Medici 'il Gottoso': Art in the Service of the Medici*, Berlin: Akademie Verlag, 1993, p. 9.

19 Rubinstein, *The Government of Florence Under the Medici*, p. 155.

20 Vespasiano da Bisticci, *Le vite*, Vol. 2, pp. 191–2.

21 '*Sollecitava questo edifitio con quanta celerità poteva, et sempre dubitava non essere a tempo*', ibid., p. 183.

22 '*Cominciata la libraria perché la sua volontà era ch'ella si facessi con ogni celerità possibile, e per danari non mancassi . . .*', ibid., p. 183. Vespasiano's comments suggest an alternative interpretation of a quotation presented by Kent, who writes: 'Piero's note that as Cosimo lay dying he "reminded me of his often expressed desire to be buried in San Lorenzo" seems somewhat superfluous, in view of his father's previous arrangements, and was presumably rhetorical', for it now seems that Cosimo may have had serious doubts as to whether his burial wishes would indeed be fulfilled. Dale Kent, *Cosimo de' Medici and the Florentine Renaissance*, New Haven, CT and London: Yale University Press, 2000, p. 377.

23 Moreni, *Continuazione*, Vol. 1, pp. 15–16; Piero Roselli and Orietta Superchi, *L'edificazione della basilica di San Lorenzo: una vicenda di importanza urbanistica*, Florence: CLUSF, Cooperativa Universitaria, 1980, p. 128; Cristina Danti, Isabella Lapi Ballerini, Pietro Ruschi and Carlo Sisi (eds), *Donatello e la Sagrestia Vecchia di San Lorenzo: Temi, studi, proposte di un cantiere di restauro*, Florence: Centro Di, 1986, p. 104; and James Beck, 'Desiderio da Settignano (and Antonio del Pollaiuolo): Problems', *Mitteilungen des Kunsthistorischen Instituts in Florenz*, 1984, vol. 28, no. 2, p. 215, note 3.

24 Herzner believes, supported by Elam, that the width of the old basilica exactly matched that of the new basilica nave, not including the chapels. He furthermore believes, supported by Morolli, that the new Brunelleschi project was originally conceived as an addition to the old basilica, and that the western part of the new basilica was in fact joined with the old basilica into one rather incongruous basilica for many years after Cosimo's death. Herzner, 'Zur Baugeschichte von San Lorenzo . . .', pp. 89–115; Elam, 'Cosimo de' Medici and San Lorenzo', p. 161; Morolli, 'Non solo Brunelleschi . . .', p. 73, Fig. 53; and Morolli, 'La "croce" di Giovanni, I "più modi" di Filippo (1422–1428)', in *San Lorenzo 1393–1993*, p. 50. This theory, however, runs contrary to the evidence presented in note 26, below.

25 See note 13. It is not likely that this passage refers only to the southern nave chapels, which were under construction in 1463, as these chapels were constructed by private individuals, not Cosimo. Roselli and Superchi, *L'edificazione. . .*, pp. 104–24.

26 The complete passage reads:

> *Reliquit amore Dei, et pro utilitate anime sue Ecclesie, et Capitulo Ecclesie S. Laurentii de Florentia dimidium urcei olei, hoc est, medietatem unius barilis, sive lagene olei quolibet et pro quolibet anno in perpetuum cum onere, quod Prior, et Capitulum dicte Ecclesie teneantur, et debeant continuo tenere unam lampadem ad Oraculum, et ante Figuram Virginis Marie pictam, et sitam in secunda columna a dextris in introitu dicte Ecclesie, sive in columna, que est in dicta Ecclesia, propinquiori, et prope januam, per quam egreditur et itur recta linea in viam Stuphe, que lampas debeat continuo cum oleo retineri accensa: in hoc conscientiam Prioris, et dicti Capituli strictissime onerando ec.*

> Out of the love of God, and for the utility of his soul, he left to the church and chapter of the church of S. Lorenzo of Florence one half pitcher of oil, that is, a half a barrel, or bottle of oil each and every year in perpetuity with the obligation that the Prior and Chapter of the said church should be held and must continually keep a light at the place of prayer and before the painted figure of the Virgin Mary, situated on the second column to the right of the entrance to the said church, or at the column, which is in said church, closer by, and near the door through which one goes out and proceeds in a straight line to Via della Stufa, which lamp should be kept continually replenished with oil: most strictly burdening the conscience of the prior and the said chapter in this matter.

Moreni, *Continuazione*, Vol. 1, p. 133, note 1. In formulating the preceding translation, I benefitted from the assistance of Caroline Elam, Caroline van Eck and Jack Wasserman. This translation corrects and supersedes my previous one in Cohen, 'Ugly Little Angels', 2007, p. 286.

27 See note 9. The detailed view of the old basilica in the Codex Rustici, c. 1445 (Figure 7.6, inset), shows a small, narrow, Romanesque-style basilica with a rear campanile and seven clerestory windows, implying two nave arcades of six columns each. The basilica thus resembled the still-extant, Romanesque-style Basilica of Ss. Apostoli in Florence. The campanile was either separated from the basilica by a small gap, the north walls being coplanar, or it touched the basilica and was slightly offset to the north. Documentary evidence indicates that the campanile stood on the site of Chapel SP 66–67. I hypothesize that Dolfini laid out the projected north wall of the new basilica (assuming deep, square nave chapels) on the location of the north wall of the old basilica, in order to be efficient in both reusing the old north foundation wall and building next to the property line. (See Figure 7.6, and Cohen, 'How much Brunelleschi?', 2008, pp. 37–43.) I also assume that the east property line encompassed the old portico. Taking the Ss. Apostoli floor plan as a stand-in for the old Basilica of San Lorenzo, inserting it into the present site plan and enlarging it to scale, while taking into account the above factors, produces the reconstruction shown in Figure 7.6, which prioritizes the Orlando testament. Note that according to this reconstruction: (1) the side door is farther east than shown in the Codex Rustici, and (2) the old portcio foundations are covered, despite possible evidence that they survived into the sixteenth century (on the latter point see Volkner Herzner, 'Zur Baugeschichte von San Lorenzo . . .', p. 112, note 50). Partial demolition of the old basilica would have been necessary prior to 1450 so that column 10 could be erected; I find no evidence that the two vertical joints in the present façade, where the chapels meet the side aisles, are in any way related to the old basilica, as Herzner suggests. Herzner, ibid., pp. 89–115; and Herzner, 'Letter to the Editor', in response to Cohen, 'How Much Brunelleschi?', *Journal of the Society of Architectural Historians*, Dec. 2008, vol. 67, no. 4, p. 112, note 50.

28 Ibid.

29 '. . . *le cappelle di nuovo simurono* . . .', Roselli and Superchi, *L'edificazione* . . ., 1980, p. 104.

30 The high clerestory walls appear to provide sufficient weight to buttress the side aisle vaults from the other side.

31 According to Brunelleschi's fifteenth-century biographer, Antonio di Tuccio Manetti, when Giovanni de' Medici was unable to assemble more than seven chapel patrons, he ordered Brunelleschi to design the basilica without nave chapels, and Brunelleschi complied '. . . unwillingly, because it seemed to him a miserable thing . . .'. Antonio Manetti, *Vita di Filippo Brunelleschi*, Giuliano Tanturli (ed.), Milan: Edizioni il Polifilo, 1976, p. 108.

32 The document also notes that the southern chapels were complete by this time. Ginori Conti, *La Basilica di S. Lorenzo* . . ., pp. 72–3.

33 The old basilica therefore must have been demolished some time between 1465 (Ibid.) and 1469. Roselli and Superchi, *L'edificazione*, 1980, pp. 128–9. The document of 1481 that Saalman claims is related to the demolition of the old campanile, I find ambiguous. Saalman, *Filippo Brunelleschi: The Buildings*, 1993, p. 439.

34 Cohen, 'Ugly Little Angels', 2007, Figs 27, 29, 33 and 34. Since the Medici construction ledger maintained between 1442 and 1453 indicates that, in the first phase of nave construction, decorative stonework was manufactured only as the new walls became ready to receive them, we may assume that this practice continued in later years. Hyman, *Fifteenth Century Florentine Studies*, Part II, pp. 325–65 and 534–47. The scenario proposed here could explain the vertical masonry joints in the façade between the side aisles and nave chapels to which Herzner refers (see note 26). The southern nave chapels could have been completed before the southern side aisle and nave arcade. Perhaps only when the central nave and northern side aisle were completed was the façade constructed in front of the central nave and both side aisles as one integral structure, thus forming the southern vertical masonry joint in question. When the northern nave chapels were later completed, the northern vertical masonry joint would have been formed. The fact that the northern joint stops halfway up the façade perhaps indicates that construction of the last northern nave chapel caught up with the construction of the main façade wall at this point.

35 Eugenio Casalini, *La SS. Annunziata di Firenze*, Vol. 1, Firenze: Convento della SS. Annunziata, 1971, p. 32, note 13, and p. 29, note 7; and Franco Borsi, Gabriele Morolli and Francesco Quinterio, *Brunelleschiani*, Rome: Officina Edizioni, 1979, p. 275. On Pagno, see also: H.W. Janson, 'Two Problems in Florentine Renaissance Sculpture', *Art Bulletin*, 1942, Vol. 24, pp. 326–9; Saalman, *Filippo Brunelleschi*, p. 194; Hyman, 'Fifteenth Century Florentine Studies . . .', pp. 139–44 and 394–7; Hyman, 'Notes and Speculations . . .', pp. 111 and 117, Documents 41–4; and Cohen, 'Ugly Little Angels', 2007, pp. 286–7.

36 On the public reception of the initial construction, see Manetti, *Vita . . .*, 1976, p. 109.

Chapter 8

Passion and control

Lewerentz and a mortar joint

Mhairi McVicar

'His passion for joints is obvious',[1] writes Bengt Edman with regard to the mortar joints at Sigurd Lewerentz's Church of St Peter in Klippan, Sweden.[2] The passion within the mortar joints at Klippan emerges from a process of conversation and risk (Figure 8.1). They remain, however, obsessively controlled. Prevailing practices of quality control in the field of architecture emphatically reject risk, prescribing certainty as a means of control, yet such practices arguably engender a quality of averages, and impede a quality of extremes. The quality of extremes – the extraordinary and shocking quality – present in the mortar joints at Klippan speak of a different process altogether.

Passion and control are both evident within the mortar joints at Klippan. The goal, observes Edman, 'is evidently not a conventional neatness [. . .] It's like baking a bread. The bricklayer is present.'[3] 'Neatness' is rejected here in favour of mortar joints that narrate progressive decisions made throughout the construction process. The joints appear rough, irregular, unfinished: seemingly uncontrolled. Varying widely in dimension – ten, fifty, one hundred millimetres and beyond – wiped and smudged across bricks, bearing rough imprints of fingers, the mortar joints inverse conventional mortar and brick relationships, are brought forth as a primary material, taking over, subsuming the brick. At Klippan, the goal is categorically not a conventional neatness; yet, clearly, it is not uncontrolled. In fact, the mortar joints at Klippan are controlled with certainty (Figure 8.2).

Rules and deviations

The obsessive control of the mortar joints at Klippan began with strict adherence to three unyielding rules established by Lewerentz, as Colin St John Wilson describes:

8.1
Brick floor at Church of St Peter, Klippan, Sweden, 1963–6. Architect: Sigurd Lewerentz (1888–1975)

8.2
Window detail at Church of St Peter, Klippan, Sweden, 1963–6. Architect: Sigurd Lewerentz (1888–1975)

> In the first place we find that the use of brick is subject to three propositions stringently applied in the teeth of commonsense compromise. First, Lewerentz proposes to use it for all purposes: wall, floor, vault, rooflight, altar, pulpit, seat. Second, he will use only the standard, full-size bricks; there will be no specially shaped bricks. Third, no brick is to be cut.[4]

There is, emphasises St John Wilson, no 'commonsense compromise'. No brick will be cut. There is to be no exception.

In standardised UK masonry construction, the mortar joint is instructed to remain dimensionally consistent. A nominal ten-millimetre joint is permitted a tolerable deviation of a millimetre, or two; the mortar joint is to remain subservient to the brick. The brick, a consistently prefabricated unit of assembly, is pre-cut and shaped to reconcile all foreseeable deviations. At Klippan, this hierarchy is reversed. At Klippan, it is the prefabricated brick that remains, without exception, dimensionally consistent. Because, according to the control of the three rules, no brick can be cut, the mortar joint must reconcile all irregularities during construction: accommodating slopes of wall and roof junctions, negotiating angular corners, mediating directional changes and masonry openings. At Klippan, the mortar joint, accommodating deviations, comes forth as a primary material. That this is permitted testifies to the engagement, care and tolerance of risk practised on site, between architect and builder, during construction.

'The decision,' writes Nicola Flora, 'to use uncut bricks for the whole building, in fact, involved the commitment to be present every day on site to resolve problems arising there, calculating, in each case, the width of the mortar joint between the courses, and allowing it to vary according to particular needs.'[5] Inviting the mortar joint to vary at all invites interpretation, and any interpretation is risky; it cannot be precisely written or drawn in advance of construction. At Klippan, this risk was negotiated on an almost daily basis through conversations between Lewerentz and the site foreman, Carl Sjöholm. 'Lewerentz slowly made up his mind',[6] notes St John Wilson, by way of these conversations. Lewerentz and Sjöholm would often work 'far into the evening, planning the next day's work on endless alterations and on site revisions'.[7] Frequent conversations between architect and builder permitted the risk of alterations and revisions during construction; on-site adjustments could respond to unpredicted circumstances and slowly developing ideas. The predictions of the drawings were overruled by conversation.

The dimensions of the mortar joints at Klippan are permitted to achieve extreme limits, structurally and visually. This construction speaks of far more than the consistent binding of bricks; these mortar joints could not be constructed with standardised repetition, inattention or thoughtlessness. Every joint, as a primary material, required careful attention and skill from the bricklayer. These unexpected, extreme mortar joints demand physical and emotional reactions; it is impossible to remain neutral when encountering the quality of extremes at Klippan (Figure 8.3).

8.3
Corner detail at Church of St Peter, Klippan, Sweden, 1963–6. Architect: Sigurd Lewerentz (1888–1975)

Specialisation and separation

Such extremities are becoming rare in current architectural practice, which is increasingly averse to the engagement and risk practised at Klippan. Direct conversations between architect and builder, prior to or during construction, are discouraged in lieu of communication methods that purport to be precise and certain: the detailed construction drawing and the written specification. Reliance upon such communications is relatively new, emerging from developments that began at the end of the nineteenth century, as Mohsen Mostafavi and David Leatherbarrow write:

> Architecture made out of a greater number of mass produced parts also changed the relationship between the architect and the builder by largely reducing the role of the latter's knowledge of traditional ways of building and relying upon construction procedures almost entirely prescribed by the architect. Independent of the architect's instructions for assembly, construction could not proceed.[8]

Integration of new, varied and complex construction methods exceeded traditional limits of construction knowledge in what Mostafavi and Leatherbarrow describe as an inversion of traditional architect/builder roles: no longer could the 'gentlemen architects' rely upon the knowledge of the builder.[9] Historically, the architect, preparing instructions for construction, could draw from the builder's accumulated knowledge and tacit skills; in a context of mutual trust and shared understandings, instructions to the builder could be limited. New, diverse and complex materials and construction methods significantly revised these expectations, even as the professionalised conversion of the architect from the on-site 'master builder' to the office-bound intellectual cemented growing divisions between architect and builder.[10] Specialisation became prevalent, and with it, a separation of knowledge. Instructions to the builder were now, increasingly, comprehensively detailed, subdivided within specialised roles, controlled by the architect. The architect in this new, inverted relationship would instruct; the builder would comply.

Subdivision of knowledge within professions is a common result, observes Donald Schön in *The Reflective Practitioner*, of the specialisation of professionals. 'Many practitioners,' he suggests, are 'locked into a view of themselves as technical experts.'[11] Professional roles are rigorously protected, Schön suggests, by preserving the appearance of knowledge. Uncertainty, he thus concludes, is seen as a threat: 'its admission is a sign of weakness.'[12] Specialised professionals, in preserving control over specialised areas of knowledge, cannot admit uncertainty; to do so may undermine the appearance of knowledge and challenge authority and control. While specialisation and separation inevitably create gaps in knowledge, all attempts to bridge these gaps, to ask for advice, to engage in conversation in order to share knowledge are potentially to admit uncertainty and relinquish control. With conversation no longer viewed as a viable means of communication, it is to the certainty of the precise instruction that practitioners are now advised to turn.

'The objective is certainty': specifying quality

Architects today are thus urged, or required, by regulatory bodies, professional affiliations and practice manuals to assure quality control through the implementation of increasingly precise instructions. In the *Architects Journal*, Francis Hall states:

> The one certain opportunity available to an architect to set down a definitive and enforceable expression of standard and quality is by way of a properly drafted specification. If this is done, there is understanding and certainty all round. If it is not, there is often disagreement and disappointment.[13]

Hall is unequivocally clear on this matter: understanding and certainty emerge from the properly written specification. Quality, in his viewpoint, cannot emerge from the unpredictability and imprecision of conversation; all details of assembly must be comprehensively predicted and clearly communicated to the builder in advance of construction. There is to be no exception. 'The objective,' emphasises Hall, 'is certainty.'[14]

The specification writer is thus bound to the onerous task of communicating, with certainty, every architectural intention to the builder. Jack Bowyer, in *Practical Specification Writing*, confirms this decisive role: the specification writer, he asserts, must be able to 'express the architect's requirements in clear, technical and precise written form, free from any ambiguity'.[15] This presents a daunting task for the specification writer; every nuance of architectural intention must be shared between architect and specification writer, and interpreted and translated into a precise language, in order that the builder may 'correctly' receive and implement precise instructions into built form. The task is daunting, but assistance is at hand. The British Standards Institution (BSI)[16] has, since the early twentieth century, responded to the growing complexity of construction materials and methods with a comprehensive and continuously updating framework of generic and specific categories of technical language. The specification writer is invited to turn to the language of BSI in order to outline and confirm virtually any aspect of the physical construction process, including that of the specification of a mortar joint.

The specification of a mortar joint

British Standard 5628 Part 3 states that, 'the joint is the medium where the variabilities due to both induced and inherent deviations can be absorbed'.[17] Elsewhere, BSI prescribes that masonry should be set out with 'nominal 10mm joints, unless otherwise specified'. [18] The masonry joint, thus prescribed, must remain dimensionally consistent – a nominal 10mm – while accommodating a wide range of predictable, unpredictable, man-made and material deviations. There is little room for error. The specification writer could be forgiven for viewing the mortar joint with some trepidation and suspicion. The mortar joint is, after all, the least certain moment in a masonry wall, as Eric Vastert warns in *Architectural Science Review*: 'The weak spots in the performance of buildings are not so much the building materials, but rather the connections between them. Dimensional variations of products become particularly clear in the joints.'[19]

The brick, as a prefabricated unit subject to factory-controlled tolerances and verifications, can be considered acceptably predictable and certain. The certainty of the mortar joint, however, remains in the hands of the bricklayer; in accumulated, tacit and personal knowledge; in skill, and care, on site. As a generic, prefabricated unit, the brick may be tolerably predicted; as an individual *in situ* construction, the mortar joint remains vulnerable. Lack of skill or care threatens to invalidate the precision of the detailed drawing or written specification. In an ideology of certainty, the mortar joint is viewed as potentially deviant and disruptive; the only guarantee of quality may lie in the validation of the mortar joint, in a final agreement between architect, specification writer and builder that the mortar joint, as constructed, meets precisely specified standards.

If the quality of the construction of the mortar joint is placed at risk through personal interpretation, the validation of the mortar joint is equally at risk. Final acceptance may rely upon the architect's opinion. Yet any process that accepts personal judgement is uncertain; if certainty is to be the objective, this cannot be accepted. 'Although', asserts Francis Hall, 'the architect's satisfaction may be a treasured thing, the question is whether it is a commodity which the contractor can reasonably cater for in his pricing and execution'.[20]

BSI thus advises that 'design outputs should be documented in terms that can be verified and validated against design input requirements'.[21] This presents another onerous task for the writer of the specification of a mortar joint, who must now predict with certainty the quantification of visual quality. BSI initially offers little support, providing a disclaimer that its specifications are 'intended to provide satisfactory structural performance [. . .] they should not be regarded as defining acceptability of appearance'.[22] If the specification writer must reject 'the architect's satisfaction', a more precise and predictable method of quantifying visual quality is required. BSI finally proposes, as a solution, the provision of an approved test panel of masonry, to be viewed next to a completed section of wall at a distance of three metres, at which distance the two panels should not differ 'significantly'.[23] Even this detailed provision is fraught with uncertainty and imprecision, leaving the term 'significantly' open to personal interpretation; the 'architect's satisfaction' may yet determine the outcome.

The 'architect's satisfaction' is unpredictable; certainty, however, 'is the objective'.

The precise quantification of visual quality

In *Architectural Science Review*, Eric Vastert is specifically concerned with this dilemma, contending:

> The problem particularly figures in connections and joints between elements where dimensional deviations have become so visible and so large that they no longer are acceptable. But when exactly is that the case? Do we draw the line at 1, 2 or 5 millimetres? Indicating the proper visual tolerance is extremely difficult for a completely visualized and dimensionalised design. The quantification of the required visual performance seems well nigh impossible.[24]

'The architect's satisfaction' threatens the ideology of certainty, relying as it does on personal judgement. Even descriptions of dissatisfaction with the quality of a mortar wall are difficult to precisely quantify. 'Façades are often felt to be ugly',[25] observes Vastert. Here, the validation of the visual quality reverts to emotional and personal judgement, and this, in the viewpoint of quality control, cannot be tolerated. As an alternative to the imprecision and unpredictability of emotional personal judgement, Vastert proposes a mathematical formula: $A - M \times C \times V$. The 'maximum permissible dimensional deviation A' is equal to the multiplication of a 'measurement factor M', a 'contrast factor C' and a joint breadth 'V'. This formula, along with the recommended use of 'reference patterns' and 'reference façades', offers to quantify the acceptability of accumulative deviations that may occur throughout a façade; 'unwelcome surprises due to the undesirable distribution of dimensional deviations', concludes Vastert, 'can be prevented' if formulaic precautions are taken.[26]

The quantification of the visual quality of a masonry wall is here taken to extreme measures, but such formulaic precision appears inevitable if the language of quality control is extended to its logical conclusion. In rejecting individual interpretation, quality control rejects the uncertainties of individual skill or care. To assure certainty in all circumstances, a lowest common denominator must be assumed; the goal must be set sufficiently low. Quality, thus viewed by BSI, is categorically not a degree of excellence, but is defined merely as 'fitness of purpose'.[27] Yet, even at the scale of the specification, construction and validation of a single mortar joint, can 'fitness of purpose' ever be considered good enough? Quality assurance, Timothy Ostler points out in the *Architects Journal*, 'will not make a bad architect a good one, but will just make him more consistent in producing bad architecture',[28] assuming the lowest common denominator may reliably predict consistency, but it does not invite or celebrate practices that contribute to the pursuit of quality as excellence: passion, care and engagement.

More than 'good enough': an ideological stance

'It matters not what the thing is', wrote John Ruskin, 'as that the builder should really love it, and enjoy it, and say so plainly'.[29] Quality control inverses this stance: it matters not at all what the builder thinks; all that matters is what the thing is, and whether the thing is predictable, certain and quantifiable. This has significant practical implications; George Atkinson, in *Guide Through Construction Quality Standards*, reports that 'lack of care was found to be a more important cause of faults on sites than lack of skill'.[30] But it is in the expectation of quality where the implications are most evident. A framework that assumes lowest common denominators cannot expect any more of its practitioners than 'good enough'; a framework that rejects conversation and engagement does not invite care to occur in practice. Rather, it anticipates scepticism, mistrust and low expectations. It is difficult to ask for more than mediocrity in such circumstances.

At Klippan, the quality of the mortar joints could not have been precisely predicted in written or drawn form; equally, they could not have been validated though generic quantification. The mortar joints achieve quality by accepting and celebrating risk and engagement throughout construction. Despite this, control is obsessively maintained at Klippan, within Lewerentz's three rules – that brick would be used throughout, that no special shapes would be used, and that no brick would be cut – rules that permitted adaptation without loss of control. These rules did not, in themselves, assure quality. They proposed an ideological argument, a working method that developed and established itself through 'endless alterations and on site revisions'.[31] Conversations between architect and builder on site were permitted to overrule predictions made by the drawings, raising the quality of the mortar joints at Klippan beyond predictably consistent construction. Conversations between architect and builder permitted the presence, knowledge and skill of the bricklayer to be intimately expressed within each mortar joint.

'What we have to admire', continued Ruskin, 'is the grand power and heart of the man in the thing, not his technical or empirical way of holding the trowel or laying the mortar'.[32] While the legalities, economies, regulations and complexities of current architectural practice and construction procedures rarely permit the engagement and risk witnessed at Klippan, it is notable that, in a drive to promise quality control through certainty, it is a quality of mediocrity that has become acceptable, through a process that arguably engenders further separation, scepticism and lack of care. In the drive to pacify the growing complexities of contemporary architectural practice, it is the simplest of human relationships that are being erased; without conversation and care, mediocrity is all that can be reasonably expected. The extraordinary quality evident in the mortar joints at Klippan speaks of a different quality control process altogether; control that begins with the certainty of passion, and risk (Figure 8.4).

8.4
Interior, Church of St Peter, Klippan, Sweden, 1963–6. Architect: Sigurd Lewerentz (1888–1975)

Notes

1 Bengt Edman, 'Lewerentz the Bricklayer', *Spazio E Societa*, 1993, vol. 16, no. 64, pp. 76–83 (p. 78).
2 St Peters Church, Klippan, Sweden, constructed 1962–6, Architect Sigurd Lewerentz (1888–1975).
3 Edman, p. 78.
4 Colin St John Wilson, 'Sigurd Lewerentz: The Sacred Buildings and the Sacred Sites', in Nicola Flora, Paolo Giardello and Gennaro Postiglione (eds), *Sigurd Lewerentz 1885–1975*, Milan: Electa Architecture, 2001, pp. 11–33 (p. 18).
5 Nicola Flora, Paolo Giardello and Gennaro Postiglione (eds), *Sigurd Lewerentz 1885–1975*, Milan: Electa Architecture, 2001, p. 336.
6 St John Wilson, p. 22.
7 St John Wilson, p. 22.
8 Mohsen Mostafavi and David Leatherbarrow, *On Weathering: The Life of Buildings in Time*, Cambridge, MA: MIT Press, 1993, p. 21.
9 Mostafavi and Leatherbarrow, p. 23.
10 For a discussion of the emergence of the professional architect in the early nineteenth century in the UK, see Mark Crinson and Jules Lubbock, *Architecture: Art or Profession? Three Hundred Years of Architectural Education in Britain*, Manchester and New York: Manchester University Press, 1994.
11 Donald M. Schön, *The Reflective Practitioner: How Professionals Think in Action*, Aldershot: Arena, 1995, p. 69.
12 Schön, p. 69.
13 Francis Hall, 'Specifying for Quality', *The Architects Journal*, 8 June 1994, vol. 199, no. 23, p. 38.
14 Hall, p. 38.
15 Jack Bowyer, *Practical Specification Writing: A Guide for Architects and Surveyors*, London: Hutchinson, 1981, p. 11.
16 'Since its foundation in 1901 as the Engineering Standards Committee, BSI Group has grown into a leading global independent business services organization. The Group now operates globally through its three divisions: BSI British Standards, BSI Management Systems and BSI Product Services.' Available at www.bsi-global.com/About-BSI/ (accessed 19 February 2009).
17 British Standards Institute, *British Standard BS 5606:1990 Guide to Accuracy in Building*, London: British Standards Institute, 1990, p. 23.
18 British Standards Institute, *British Standard BS 5628–3:2005 Code of Practice for the Use of Masonry – Part 3: Materials and Components, Design and Workmanship*, London: British Standards Institute, 2005, p. 102.
19 Eric Vastert, 'Specification of Visual Requirements for Regularity of Joints', *Architectural Science Review*, September 1998, vol. 41, pp. 99–104 (p. 99).
20 Hall, p. 38.
21 British Standards Institute, *British Standard BS ISO 9000–2:1997 Quality Management and Quality Assurance Standards – Part 2: Generic Guidelines for the Application of ISO 9001, ISO 9002 and ISO 9003*, London: British Standards Institute, 1997, p. 6.
22 British Standards Institute, *British Standard BS 5628–3:2005 Code of Practice for the Use of Masonry*, p. 101.
23 British Standards Institute, *British Standard BS 5628–3:2005 Code of Practice for the Use of Masonry*, p. 126.
24 Vastert, p. 100.
25 Vastert, p. 100.
26 Vastert, p. 101.
27 British Standards Institute, *British Standard BS 4778–2:1991, Quality Vocabulary – Part 2: Quality Concepts and Related Definitions*, London: British Standards Institute, 1991, p. 3.
28 Timothy Ostler, 'Does Quality Assurance Work?', *The Architects Journal*, February 1994, vol. 16, p. 32.
29 John Ruskin, *The Stones of Venice*, edited and abridged by J.G. Links, London: Pallas Athene in conjunction with Ostara Publishing, 2001, p. 23.
30 George Atkinson, *Guide Through Quality Control Standards*, Wokingham: Van Nostrand Reinhold, 1986, p. 19.
31 St John Wilson, p. 22.
32 Ruskin, p. 21.

Part 3

Belief

Quality will out. It will, in the end, make itself apparent. For many, this is a matter of belief. Even if careful control of construction does not reliably predict quality, even if the mechanisms of its appreciation shift and mutate, the aura of quality will shine through.

In Chapter 9, David Leatherbarrow questions the conditions in which quality comes into being. He argues that architectural qualities emerge before, during and after construction. They are ascribed by external agencies – through processes that he calls qualification – and follow from the properties of materials, their finishing and the physical traces of inhabitation that mark them. By undergoing qualification and requalification, he believes, elements can come to obtain high quality, suitability and character.

Marc Treib, in Chapter 10, believes that quality inheres in simplicity. He likens contemporary architectural production to blockbuster film production, arguing that the stories remain the same but that 'viewer interest can only be maintained by making the effects more spectacular with each retelling of essentially the same story'. Challenging this fashion-conscious, image-driven view of architecture, he argues that quality instead emerges from elegant solutions and simple things. Profundity is to be found in simplicity, Treib suggests, thought of as condensation rather than reduction.

There are plenty who have found quality emergent in the architecture of James Stirling. Igea Troiani, in Chapter 11, reminds us that his early brick buildings (some designed in conjunction with James Gowan) became sites of pilgrimage for the faithful followers of modern architecture because of their formal innovations and their manipulation of surface and material. Troiani examines Stirling's troubled Florey Building for Queen's College Oxford, finished in 1971. She contrasts the belief in the architectural qualities of the student residence building held by commentators with the opinion of the client, who eventually sued the Stirling office because of problems with privacy and environmental performance and because of overruns of time and cost. Quality, here, is in the heart of the believer.

Chapter 9

Necessary qualifications

Design before, during and after construction

David Leatherbarrow

In architecture today, there is considerable discrepancy between designers' ability to generate alternatives and their ability to judge which among them has the highest quality. The problem gets worse when shared judgements are sought. Indefinite and subjective as design judgement seems to be these days, there was a time when architects believed that the quality of a work was self-evident, not only to the designer, but to all concerned. Claude Perrault, for example, listed the richness of materials as one of architecture's 'positive beauties' because such assessments are made 'naturally and by necessity'. A marble building, he suggested, has more quality than one made from timber. Even if such a claim would receive little support today, one wants to believe that personal taste is not the only alternative. Instead of assuming that quality is a matter of common sense, as Perrault did, or thinking it is largely personal, as we tend to, I want to pose a more basic question: under what conditions does quality come into being? When does it not? How might we account for its lack or supply? To answer these questions I will argue three points: (1) that quality results from different kinds of qualification; (2) that many 'agents' have a hand in its emergence; and (3) that these agents impart quality over time, not always according to schedule and expectation. Although these points may seem minor, I hope to show they point toward a new conception of architectural design, one that follows a longer schedule, and has a narrower scope and a much richer set of sources.

The English word quality has equivalent forms in French, Italian and German: *qualité*, *qualità* and *qualität*. These terms and their cognates in other European languages derive from the Latin root *qualitas*, meaning relative kind, character or nature. English also has an obsolete but indicative cognate, quiddity, the adjectival form of *quid*, meaning that which makes a thing what it is. Alberti's famous motto *quid tum* asked 'what then?' or 'what next?'. Cicero used these Latin words to translate Greek vocabulary. In Homeric usage, *poios* meant 'of what nature or kind'. That not only distinction but also appearance was implied is clear when one recalls that the related Greek word for making, *poiesis*, indicated something wrought *and* apparent, because everything made was made *visible*. Assessments of high or low quality are generally thought to depend on judgement or taste. We often assume this is either personal or the intelligence of connoisseurs who possess special powers of discernment. But my brief etymological review shows that qualities can also be seen as the subject matter of everyday perception, insofar as they make commonplace things evident, or are what each of us hears, feels, or sees when we work and live with the things and places of the world – the tones, tastes and traits of all that surrounds us. Quality, then, is the 'what' of something, the content of its character, and it is visibly manifest. Qualities also individuate things. They are the attributes or properties of a person or place, possessed by it alone. Thus, any question about the conditions under which quality emerges must take up both appearance and distinction, no matter whether they result from self-actualisation (as with people) or cultivation (as with things).

The link between something's identity and perceptible attributes has been the subject of great debate in the history of philosophy. There may be no more familiar topic in the long history of epistemology, nor a theme that is so central in the phenomenology of perception. When described as accidents in metaphysical thought, qualities were taken to be contrary to substance. Yet, when moral philosophers argued that qualities expressed character, they were thought to show what is true, natural or unchanging about a person or thing. Theophrastus, Aristotle's successor at the Lyceum, held this second view. He thought that studies of character would help a young person learn which types to avoid and which to imitate. For him and many to follow, such as Le Bruyére and Shaftesbury, characteristics indicated the true nature of a person. In our time, too, characterization serves as a way of arriving at both distinctions and rankings. Some characters are high, others low, making them admirable or avoidable. Derivative practices, such as caricature, stereotyping, profiling and branding, also show high and low quality, but shortcuts such as these prompt the old question concerning the superficiality of qualitative judgements. No matter whether one sees quality as substantial or accidental, it is always *seen*. My concern, again, is with the conditions of that appearance.

In architecture, two topics can be seen to largely condition a setting's quality: *finishing*, or the surface treatment of construction materials, and *configuration*, or the relationships among an ensemble's several parts. Identifying these topics as decisive does not mean others are unimportant. Because a building must remain standing to be perceived, structural integrity is required for quality to be apparent. Yet, the role of building structure in qualification is indirect. Structural behaviour is inferred on the basis of apparent qualities. My next point is key: while seen, quality is never apparent 'in itself'; inherent qualities are not what settings show.

Some of the synonyms of the word's verb form – to qualify – point to the *circumstantial* or *regional* premises of qualification. Dictionaries explain that to qualify is to temper, adapt, adjust or conform. To conform is to bring something or oneself into accord with some pre-existing pattern or principle. Adjusting is slightly different: it results in a close correspondence between one thing and the others with which it has contact, the way a camera lens sharpens the image of an object that is near or far. To adapt means to modify something so that it meets new conditions, an old speech to a new audience, for example. None of these modifications happens in isolation. Each assumes some context in which qualification occurs. The changes that result show the influence of those circumstances. Think of foreigners moving to a new city: once acclimated, they show themselves to be qualified to participate in its typical situations. Just as qualification is situated, so is character: no single element defines a street's character; all of its parts in their precise interrelationships play a role. In both character and quality, *interaction* is essential.

Like any process, qualification unfolds in time. The terms I just introduced (adaptation and adjustment) suggest that this happens after the fact of constitution: first, something exists in itself, after which it is qualified by some external agency. Yet, this truism cannot be correct if the need for adjustment is apparent, for that, like any appearance, requires qualities – qualities that must have been formed at some point. This means adaptation always precedes itself. Conforming is always reforming. Moreover, this two-step process is followed by a third. In fact, qualification is continual. All that distinguishes the most recent from earlier steps is the force responsible for the change. The architectural examples I shall describe will show that the qualification of a setting writes a history of several chapters, narrating the labours of different protagonists. Three have been mentioned already: nature, art and accident. Although these names will quickly reveal themselves to be inadequate, at this point they should be sufficient to identify the important players in the phases through which architectural quality becomes manifest.

Three times are also key: the time before the work is undertaken, when its elements are *pre-qualified*; the time of construction, when those same elements are *qualified* through the several techniques of assembly and finishing; and, finally, the time after the work is finished, when its parts are *re-qualified* through the effects of the environment and practical use. Although these stages are variously apparent in all works, I shall exemplify them by returning to a single building, Le Corbusier's Carpenter Center at Harvard, the details of which will guide my description of the stages of qualification and be used to introduce other examples.

Pre-qualification

Because every work is predefined by the elements brought to the site, qualification occurs before the inauguration of site labour. Nothing used in construction is without qualities. Stated positively, materials, components and systems are selected on the basis of attributes they possess – properties that make them pre-qualified for use because of their pre-finishing. Generally speaking, there are two agents responsible for these preliminaries: natural and industrial processes. These days, manual labour is a less common source. I will not try to explain the processes under which pre-qualification occurs – that is the task of the natural and social sciences – only that this process does occur, that finishing precedes design and construction. The term 'working with materials' invoked by so many architects, especially since the nineteenth century, takes for granted this earlier development, in order to urge that it be continued in site labour. From Ruskin, to Wright, Kahn and Zumthor, nothing could be worse than working against the nature of materials. Yet, this concept, like quality, is far from clear. What is more, as a remnant of old-style hylomorphism, it often impedes genuine creativity.

Le Corbusier used different kinds of glazing at the Carpenter Center, and one type is remarkable for its ambiguous qualities. Glass block or brick admits light, as does all glazing, but diffuses it. Likewise for the view: it can be seen through, but barely, as if it were not glass but shallow water. Although it comes in blocks or bricks, like masonry, glass block is far more vulnerable to cracking and chipping. It shines, but unlike the reflections seen on the surfaces of plate glass, its images are broken, like a drawing on a crumpled sheet of paper. All of these qualities were well known to Le Corbusier by the 1960s, the time of the Carpenter Center's construction. He knew of Perret's use of glass block at rue Franklin in 1902. He himself had used it in the Salvation Army building, also in his Clarté building, his own apartment, and Maison de Weekend. While Chareau's use of so-called Nevada glass in Maison de Verre gave it great prominence, Le Corbusier thought glass block was best suited to light-wells or circulation spaces, where its muted glow and abbreviated view contributed to the right sort of character. None of these qualities resulted from construction finishing; they were given to the material before site work began, through pre-qualification.

Just as glass walls can be relatively opaque, those made from stone can be somewhat transparent, at least when stone is caged, for its faceted hardness prevents close packing in 'gabion' construction. The gaps in the walls that enclose the headquarters of the Dominus Winery by Herzog and de Meuron admit both light and partial views, but also air (Figure 9.1). When seen obliquely, the cage geometry restores the sense of solid continuity, even mass, by surprisingly delicate means. Yet, when seen frontally, the walls appear to be filters as much as barriers. Stone and steel offered exactly the qualities that were required of this sort of building's edges. A similarly corrugated surface results from the bricks that Eero Saarinen chose for the Hill House dormitory at the University of Pennsylvania. Viewed frontally, the skin is blotchy. When seen as fabric, it exemplifies the 'Anatolian rug' imagery Louis Sullivan recommended for brick. The diagonal view from the sidewalk, however, emphasizes the wall's massive thickness. Still, the surface has great animation, owing to the play of light and shadow across each brick's broken face. That aggregate composition does not necessarily result in the qualities I have just described is apparent in the retaining walls that define the upper

9.1
Herzog & de Meuron, Dominus Winery, Yountville, California, 1996–8

terraces of the Neurosciences Institute in La Jolla, California, by Tod Williams and Billie Tsien. The concrete appears to be perfectly smooth when seen obliquely. But, on closer inspection, the wall is amazingly composite, revealing not so much the imprint of form-work as its highly composite make-up. Likewise for the Texas fossil stone. Both materials also show their chromatic affinities with others in the building, as well as the landscape that surrounds it. Again, all things brought to a site are suitable for one or another application, because they have been pre-qualified with specific characteristics. This is vividly true when timber is used. The wood Kahn used for the study carrels in Exeter Library has qualities that confer upon the setting the right luminosity, texture, and temperature. Put in reverse, wood has the capacity to absorb the ambient qualities that are right for the situation.

The fact that *single* materials can be variously pre-finished provides design with a range of qualities to select and use. Consider glass once more. The laboratories of the Neurosciences Institute are enclosed with vertical panels of glass, each three-part. The bottom section, perfectly transparent, admits an unrestricted view, even though it has been treated to reduce glare. Above this is sand-blasted glass that has been left sufficiently translucent to illuminate the depth of the plan. The third section admits neither view nor illumination; its job is to restrict movement at the edge of the upper terrace. In Ada Karmi's recently finished Bet Avi Institute in Tel Aviv, a similar range of qualities results from the co-ordination of different types of glass in single apertures: transparent, fritted and sandwich panels, the latter serving as luminous sun screens, used vertically or horizontally depending on solar orientation (Figure 9.2). But it is not only glass that arrives to a construction site with varying qualities. Each of the matte-finish stainless-steel panels that

9.2
Ada Karmi, Avi-Chai Center, Jerusalem, 2008

encase Kahn's British Art Center in New Haven varies slightly from the others nearby. Some variations result from environmental influence, but differences also depend on the ways the panels were pre-qualified in industrial manufacture. When the façade as a whole is seen, the variations in texture relieve the surface of the monotony it might have possessed with greater uniformity of finishing. Kahn seems to have intended both consistency and unevenness, as the corners show particularly clearly. Subtle variation can also result from the internal make-up of materials. When Brinkman and van der Vlugt specified the interior finishes of their Van Nelle factory buildings in Rotterdam, they added ground-up quartz to the plaster so that it would reflect natural light deeper into the building, improving the work environment. Neutra used mica additives to brighten the walls of his Kaufmann House in Palm Springs. In all of these cases, quality resulted from the internal constitution of the building's component parts.

Colour is perhaps the most obvious and effective example of pre-qualification. In some materials, what appears exists by nature. In others, the manufacturing process is decisive. A third case is when pigment is added to the surface of less expressive materials, as in tile or glazed brick. A very well-known example of the first kind of colour qualification is Mies van der Rohe's use of onyx and ebony in the Tugendhat Villa. The colour will never change with more or less polishing, for it exists in the depth of the material. We tend to call a quality such as this 'natural', because it continually re-emerges with each act of finishing. But colour can also exist in the depth of artificial materials. A stunning example of spatial character being largely determined by elements with colour added during their manufacture is the Guthrie Theater in Minnesota by Jean Nouvel. As one moves from room to room, one passes from colour to colour, each chosen to qualify the place. The distinct hues are exceedingly intense in their qualities, no matter whether one is in a circulation space, a theatre or a lounge. Quality conferred by colour tile is different. In Skirkanich Hall by Williams and Tsien, no figures appear, but colour baked onto the bricks gives an otherwise insistently flat surface great richness and depth.

Qualification

The second moment in the development of architectural quality is *qualification*. Finishing is a more common expression, indicating not only the procedures of cutting, smoothing, and polishing, but also the results of assembly technique. With respect to both finishing and assembly, a primary distinction exists between wet and dry elements and procedures. Many of the qualities of the Carpenter Center were determined by ways of treating wet materials, mostly concrete. As with Le Corbusier's use of glass blocks, here I touch on a topic he concerned himself with for many years. I will not attempt to say anything comprehensive or synthetic about this complicated issue, but only observe that this building is no less specific in its handling of concrete than others from the post-war period – La Tourette, Chandigarh, the Unités and so on. But it is not only wet materials that are qualified in architectural construction; dry materials too are subject to alteration or adjustment. In these cases, as with the selection of pre-qualified materials, the topic to which the reworking is oriented is some specific situation, combining site and programme.

In the United States, clapboard has been the preferred cladding material for residential construction for a long time. In recent years, vinyl often replaces wood. Earlier, though, clapboard construction reached great refinement, among the Shakers for example (Figure 9.3). Like so many aspects of Shaker culture, building practice followed the principle of simplicity: the thin upper edge of each long board was nailed to an under-lament, then lapped or covered by the thicker edge of the next. Obviously, this is dry construction, the finished qualities of which resulted from the ways the boards were first planed, then spaced and finally lapped. That similarly simple practices have neither disappeared in our time nor are exclusively American is clear when one turns to the work of Tadao Ando. Normally he is cited for his handling of concrete. But Ando also knows how to handle timber, or more exactly, how cutting techniques qualify it. Upon first entering his famous Church of Light in Osaka, one focuses on the cross in the end wall. Much can be said about what one sees or does not. The symbol seems immaterial, or nearly, almost nothing. The flooring is not. It is made out of rough sawn timber and painted pitch black. Its qualities make those of the end wall apparent by contrast. Sawing also qualifies settings when stone, not timber, is used. Consider the case of Peter Zumthor's Thermal Baths in Vals, Switzerland. Inside and out, the building's walls are made of thinly sliced slabs of stacked stone, revealing at their edges not only tonal variation because of slightly different make-up, but a range of textures revealed by cutting. While consistent, the walls are wonderfully varied, the quality vivid.

Shifting from dry to wet materials, one might say that concrete finishing is the most obvious way that the character of a setting is determined by on-site

9.3
Pleasant Hill, Shaker Village, Kentucky

qualification, at least in modern architecture. If anything is on show in the open space at the courtyard of Kahn's celebrated Salk Institute, it is the interplay of sun, sky and sea, performing as if on a solar stage. Also at play, providing reversals to all that seems stable, are the shadows. The props of this performance are few, mostly concrete walls. When the entire space is viewed, their unanimated greyness seems most significant, as though they had been prepared to approximate the uniformity of cloudless sky and calm sea. But like most good buildings, these walls offer different content at different distances. When seen nearby, the geometry of the form-work subordinates the greyness, as if the scale set by the dimensions of the grid of lines and dots were the wall's primary expression. At a still closer view, outlines recede in favour of the concrete's dermal properties, seemingly sagging, freckled and porous.

Different construction techniques would have yielded different qualities. Consider, for comparison, Le Corbusier's La Tourette, also a virtuoso performance of concrete finishing, but rough. One entry to the chapel is from a descending walk that begins in the central atrium, a room that also adjoins the refectory. Several surfaces frame the threshold: the shiny spread of the floor, the rough edges of the *ondulatoires*, the slabs and columns imprinted with marks of wooden shuttering, and the truly remarkable surfaces of the enclosing walls. If the latter were compared to fabric, the material would be burlap, not silk. If something in nature, it would be scruffy terrain, such as underbrush. If skin, a few days unshaved. Yet, the texture quiets and warms the space. Both qualities define the mid-point between the dining room and the chapel, giving it the right character.

My third and last example of concrete finishing has been alluded to already: Ando's Church of Light. Here, all that Le Corbusier attempted to achieve through articulation has been silenced. Although a range of contrasting qualities is shown outside, once entry has been accomplished, nothing of that variety appears, nearly nothing is shown. Paradoxical as it may seem, the less anything appears, the more the space accomplishes its purpose. Even traces of the work's history are effaced (the sequence of several pours). Emptiness is a positive quality when the situation requires receptivity.

Wet and dry materials need not exist separately; in fact, they often occur together, as in brickwork. The qualities of a wall, and of the space it encloses, result in part from the selection of materials (brick and the ingredients of the mortar), but also from the techniques of assembly. A wonderful case in point is Lewerentz's St Peter's Church in Klippan (see Chapter 8). Lewerentz is known to have taken great care with the brickwork, despite the fact that the pattern one sees inside and out has striking irregularities. In fact, he set a number of rules for the bricklayers, the most important of which was that no brick could be cut. This meant mortar width had to change in the running length of a wall, likewise for its height. The result is at once uniform and irregular, similar, he said, to the trees that surround the building. But joints need not be expressed to be expressive. The best case to show this is the façade of Alberti's Palazzo Rucellai in Florence. The geometry of its surface is highly controlled. First, there is the sizing and spacing of the apertures, as they align with the primary grid of pilasters and entablatures. Then, there is the pattern of rustication, uniform throughout. Yet, while one would expect the outlines of the rustication to coincide with the edges of the sandstone blocks, a close-up view – of the pilasters especially – shows that actual joints are suppressed, meaning the geometries of construction and expression are misaligned. A similar reversal of expectations can be found in a twentieth-century example that uses timber, not stone: Gropius' house in Lincoln, Massachusetts. Arriving to the US from Germany, he was alert to the tradition of timber building. Yet, dedicated as he was to the image of 'the modern', he modified the traditional method and used clapboard vertically, not horizontally, approximating the unbroken planarity of European stucco. A counter example to these last two is the Postsparkasse in Vienna by Wagner.

In this case, the 'joints', the fasteners, form a pattern that gives the building wholeness and scale, as did the inscribed lines and angles on Palazzo Rucellai. The play of points in Wagner's building can be seen inside and out, regardless of the materials being attached to the structural frame.

The last category of qualifying through assembly and finishing I want to address is no doubt the most familiar: polishing. This is the normal way dry materials are qualified. As with casting, the range of treatments is particularly clear when a single material is used. Consider again Ada Karmi's Bet Avi Institute. There is no more familiar building material in Israel than Jerusalem stone. Nor are the several finishing techniques unfamiliar: quarry sawing, chiselling, bush hammering, sand-blasting and polishing. What is distinctive about this case is the precise co-ordination of these forms of treatment with one another and the several requirements of the building's distinct settings: for example, the bush-hammered finish outside the entry door and the sand-blasted finish inside – one material, two finishes, outside and in; thus, two characters, approach and arrival. Finishing aims at qualities that suit the given situation.

Re-qualification

I have said that qualification has three phases. The third, to which I now turn, is *re-qualification*. This occurs when the elements that constitute a setting are refinished by powers or influences outside the architect's or builder's control, once the work opens itself to both use and misuse by the people who inhabit it. Also influential, perhaps more so, are the effects of the natural environment.

After the Carpenter Center was built, José Luis Sert, who had collaborated on the project, recommended that the interiors be painted, because the concrete finish was uneven. He suggested light grey. As a result of nearly endless debates about both colour and technique, little painting was carried out. Yet, in the decades that followed, the concrete has become qualified, both in and outside the building, regardless of the fact that little was specified by the architects. This re-qualification is hardly unusual. More importantly, in the long history of most buildings, it is decisive for character.

Consider the reshaping of profiles through frequency of use, the 'dishing' of stone treads, for example. On buildings that are centuries old, this refinishing radically alters original profiles. Yet, insofar as these changes result from preferred movements, they could be said to improve or enrich the original solution. They certainly confer orientation by guiding movement along contours others have established – albeit unintentionally and anonymously. Modifications such as this have an especially human voice, perhaps because of the fact that not just one but many have shaped them. Speaking not of the foot but the hand, the Greek architect Aris Konstantinidis suggested these surfaces acquire a human-like skin that makes the building seem like a living organism – which plainly it is not. The door into Peter Zumthor's chapel at Sumvigt has such a 'skin', unintended in design but 'constructed' through continual use.

The section of the door below this handle reveals something different, not the re-qualification that results from contact between the building and the human body but that from its encounter with the forces of the natural world, forces that can be enriching or corrosive. Building materials work for a living. Their job is to variously admit or resist ambient forces, wind, rain, sunshine, temperature change, humidity and so on. The surfaces we see record success and failure in this work. Below the metal plate that holds the handle and contains the lock, the timber of Zumthor's door has been darkened by deposits from the metal above. The cladding, made out of larch shingle, also shows variation, not only piece by piece, but in large sections. On the north, the shingle has become dark grey in colour; on the sun side, it has retained its reddish-brown hue. These changes do not detract from the building's quality, but in fact enrich it by showing its receptiveness to the characteristics of the place, as if its epitome.

Other sorts of re-qualification pose different sorts of question, especially the phenomenon of staining. Are the deposits of soot that mark the stone of a Hawksmoor church in London evidence of the building's congeniality with its location, or do they detract from its intended qualities by confusing the basic geometry? The same question can be asked of more recent buildings: Marcel Breuer's Bijenkorf Department Store in Rotterdam, for example, where air-borne deposits have variously stained each of the façade's travertine stones, replacing a grid of perpendiculars on one side and hexagons on the other with a seemingly random spread of streaks. If the superimposed pattern gives voice to ambient conditions, does it thereby silence what the architect intended? Of all the elements that re-qualify buildings, water is no doubt the most powerful. Alberti wrote that it is architecture's greatest enemy, a foe that is eventually victorious in every single case. The fountain in the forecourt of Paul Cret's Rodin Museum in Philadelphia (Figure 9.4) was beautifully proportioned and detailed when first built. Since then,

9.4
Paul Cret, Rodin Museum, Philadelphia, 1926

the pipe that supplies water to the fountain has also conducted iron deposits to the stone surfaces, greatly modifying its designed appearance. Should the stone be cleaned, to restore the image Cret intended? Would that deprive it of the qualities by which it is now known?

The same questions can be asked of other examples of environmental influence: bleaching, saturation and oxidation. Timber surfaces show bleaching quickly and clearly. The Lewis Glucksman Gallery at University College Cork, designed by O'Donnell and Tuomey, is clad in unfinished angelim boards. Since the time of construction, some pieces have darkened and others bleached, depending on how little or much sun they have received through the surrounding trees. The change in colour of the wooden panels in Kahn's Salk Institute is more regular: those at the lower levels retain their colour longer than those at the top, where bleaching has made them almost the same hue as the concrete walls into which they are set. Yet they also vary from side to side, as a consequence of direct sun or continual shade. The building mirrors its location by absorbing the influences of ambient forces. Another term for absorption is saturation, although it usually implies that a surface is fully soaked and can absorb no more. In cases such as this, the qualities of the original surface disappear. Think of some element in the landscape, a tree or a wall, so covered with moss or lichen that the original surface is no longer apparent. The sun screen on the face of Le Corbusier's Secretariat building in Chandigarh has been so completely covered with planting that its intended appearance is now hard to recall. Likewise, its function: intended as an instrument that would resist natural forces, it now welcomes them. Saturation is also evident in Scarpa's Brion cemetery, where the obvious distinctions between elements that are wet and dry, inert and dynamic or light and dark are no longer so apparent, insofar as the environment has reclaimed elements once taken from it. Oxidation, like saturation, radically re-qualifies architectural elements. This process is clearly evident in buildings made out of metal, copper in the case of Steven Holl's addition to the University of Minnesota's architecture school, and cor-ten steel in the case of Brigitte Shim's weathering steel house in Toronto. As with alterations to timber surfaces, change in Holl's metal panels attests to the influence of wider conditions, those at ground level as opposed to those at the roof, and on the building's different sides. The pillow-shaped profile of each panel affects how it suffers environmental influence. The changes to the steel that covers Shim's building are less varied but no less dramatic, especially when seen up close, where the chromatic range is spectacular.

Perhaps no topic is given quite so much attention in current discourse than my last type of re-qualification, the brightening, shadow projection, reflection and mirroring that result from lighting. One of the most recent and accomplished examples of the play of these effects is Jean Nouvel's Branly Museum in Paris. It is a perfectly good example of what he meant by 'putting nature to work', for the qualities of the glass screen that divides the front garden from the walk along the Seine reflect fragments of both landscapes, just as its shine mirrors the drift of clouds and the light of the sun. Shadow projection can have an equally strong effect on the qualities of a setting. This effect is apparent in the so-called gloriette on the upper floor of Neutra's Kaufmann house in Palm Springs. Under the late-afternoon sun, the shadows are so strong that the primary configuration and orientation of the space are reorganized, in line with the projected pattern. Intense light can have the same effect as cast shadow. I have already described the range of surfaces in Le Corbusier's La Tourette. A beam of light descending along the line of a stair cancels all the differences the form-work established (Figure 9.5). Perhaps there is no more vivid case of environmental conditions re-qualifying what site labour had established.

Necessary qualifications

I have argued that quality requires qualification, and that the latter unfolds in stages: pre-qualification, qualification and re-qualification. Although I have presented these three in sequence for the sake of clarity, the truth is they overlap or *interlapse* – the same forces that pre-qualify also re-qualify. The same is true in lived as opposed to conceived time: the past is a former future and a recent present, the present is an impending past and a recent future, and the future is a present and past yet to come. This interlapsing occurs because each dimension of time is aimed at one other than itself. But qualification, in addition to being temporal, is regional. Qualities

9.5
Le Corbusier, La Tourette, Eveux-sur-l'Abresle, 1957–60

do not exist in themselves, they emerge in specific situations: of formation (natural or artificial), of construction (wet or dry), of use (planned or unforeseen) and of acclimation (illumination, oxidation, saturation, bleaching, and staining). I called this interaction and stressed mutualities. This point is difficult because of our dogmatic commitment to 'the things themselves'. The interactions between a given element and the conditions of its appearance disclose the tendencies or potentialities of each. The element and its situation are correspondingly indicative. This is why qualities do not exist in themselves – they exist in the world – and why they change over time.

If all things have qualities, why do we see some as high and others low? How are those determinations made? With respect to what scale or measures are rankings decided? When is an element over- or under-qualified? Are estimates of high and low, under and over based on 'intrinsic' worth? That cannot be so if my arguments about 'external' agencies of qualification are allowed to stand. Instead of having absolute value, these estimations are based on the nature and requirements of the situation in which they exist, in which qualification occurs. To be under- or over-qualified means that the element is more or less refined than is appropriate to the situation in which it appears. An example might be stone, finished like glass, appearing at a building's base, or *beton brut* used for bookshelves, or glass block in a window facing a beautiful prospect. One cannot say any one of these materials is *necessarily* under- or over-qualified – too smooth, rough, or translucent. Judgements of quality in architecture are no less conditional than the things being judged. In fact, a number of the wonderful settings I have described obtain their quality through elements that initially seemed out of place. Through both qualification and re-qualification, elements *come to obtain* high quality, also suitability and character, because quality is not intrinsic to things. If anything is intrinsic, it is the *capacity* for qualification. Well-qualified means that an element has been prepared to *co-operate* in the realization of a setting's character, no matter whether that qualification occurred before, during or after construction.

Chapter 10

Simplicity and belief

Marc Treib

There seems to be little question that we live in an age of complexity, perhaps undue complexity. Internet access, the interrelation of almost all the people all the time, the pressures of urbanisation, natural resource depletion, increasing traffic and pollution, and our ability to be anywhere at any time electronically – all have led to a rather complex living mode indeed. Architects, heeding the call, have produced an architecture correspondingly complex. Falling to architectural fashion trends, market pressures and the surge of corporate capitalism, architects now give their clients just what they want, or what they themselves want to give them. It is a world of special effects and, like the action movie, viewer interest can be maintained only by making the effects more spectacular with each retelling of essentially the same story. It is an era of bigness, as certain architects continue to tell us, an era of outright commercialisation. And if market forces suggest that we should engage in complexity, fragmentation and the curvilinear – well, why not? But could we not – should we not – think in a somewhat different and more responsible way? Rather than *following* the market forces and the increasing complexity of life, should we not *resist* them?[1] Rather than the MP3 moment of my-music-all-the-time, what about a moment of silence, a moment disengaged from entertainment?

Silence

As far back as 1978, the Japanese architect Takefumi Aida wrote that the architect's gift to contemporary life was to provide a moment of silence in a tumultuous world.[2] In Aida's work of the 1970s, we see a conscious rejection of the notion of style, a turning away from the social sphere and the street and a turning inward towards a world of asymmetrical dissonance or curiously formal symmetries (Figure 10.1). Titles such as Annihilation House or Nirvana House lucidly revealed the architect's intentions, that is, to fashion a small world that, in its limited ways, rejected the urban sphere around it and, in other ways, created a realm not that far from it. After all, no style is a style in itself.

10.1
Takefumi Aida, Nirvana House, Tokyo, Japan, 1972

Tadao Ando's early houses relied on a similar stance, cutting themselves off from tradition as well as from their urban contexts, looking to the sky rather than to the neighbours.[3] Some architects still provide us with architectural moments such as these; others acknowledge the presence of the natural world; and others still acknowledge the importance of people to architecture. Ando is one of them; John Pawson is another; the Finnish architect Aarno Ruusuvuori was a third.[4] But here we enter the dangerous ground of the *minimal*, which needs be distinguished from the *quiet*. Minimal tends to discard rather than embrace. Quiet, in contrast, admits complexity – but only a complexity that is modulated, that is distilled and condensed.

Simplicity

I confess to having a fondness for elegant solutions and simple things. This fondness is due perhaps to a modernist architectural education, perhaps to values acquired from parents who had lived through the Great Depression of the 1930s. Whatever the reason, I am troubled by gratuitous complexity, with forms created solely for form's sake – especially if they are conceived at the expense of space and utility. I find vapid the complexity that denies the richness of human experience and the tradition of social institutions, or an architecture that confuses complexity-in-form for complexity-in-experience. Simplicity of form, complexity of experience, economy of means.

Simplicity should not be equated with reduction per se. In the West, we conceive of simplicity as a state acquired by leaving things out. Instead, could we not think of simplicity as the product of putting things *in*? Rather than overlaying them, however, or juxtaposing them as identifiable fragments, suppose we compress a multitude of aspects into simple forms – forms that read as simple although incorporating a world in themselves, like the residue of past eras trapped in amber?

Nature teaches us about the economy of means. The species that survive usually have been pared down – through evolution – to allow for maximum efficacy in negotiating their environments. Even the bulkiest of animals – the whale and the polar bear, for example – endure by possessing appropriate skeletal structures, skins or furs that allow them to thrive in cold waters. Bone structure provides a lucid illustration of this economy: the skeleton or a carapace has been honed to form a well-distributed, minimal mass that efficiently supports the weight of the animal. The bone of the terrestrial elephant is dense and large because it supports tons of weight. The avian bone, in contrast, is a model of lightness, with a hollowed structure – like a truss or space frame – a lightness that permits flight. Crab and lobster shells derive their strength not only from their respective chemical compositions but also from the folding that stiffens a thin sheet – a model engineers such as Felix Candela and Eduardo Torroja adopted in the last century for the construction of concrete shell structures. At the turn of the twentieth century, D'Arcy Wentworth Thompson investigated the world of the microscopic animal and found in the diatom a similar structural logic.[5] One could also cite examples from the floral world: seed-pods, stems, leaves and even petals as effective vehicles employing nearly minimal means. All of these examples illustrate a simplicity in structural form and a trend towards minimal amounts of material.

Biologists also talk of a 'target of opportunity', the profitable use by animals of a particular situation. For example, the bowerbird appropriates anything within his range to decorate the bower by which he attracts a mate. In the wild, his materials may be leaves, grasses and adorning berries; nearby cities, he cleverly utilises the paper wrappers from drinking straws and even an aluminum camper's cup, nicked for its sparkle.[6] One profits from what the situation provides, applying that material in an efficient way.

One wonders, in contrast, about the structural efficiency of buildings, and even bridges, by Santiago Calatrava. While I claim no expertise in structural engineering, this bridge over a very narrow canal in the Netherlands appears to be rather overblown (Figure 10.2).[7] In contrast, the bridges of the Swiss engineer Robert Maillart seem of a higher order. Maillart maximised the potentials of his material – usually concrete – and the building techniques then available to him to create bridges of beauty and even wonder, bridges that were as stimulating to traverse as they were to view from a distance.

10.2
Santiago Calatrava, canal-bridge, Haarlemeer Polder, Netherlands, 2004

Effigy

The Shinto religion of Japan is among a very small number of belief systems that possess neither an iconography nor a holy book.[8] In most religions, humans – or demi-humans – have acquired the status of godhead, or the gods have been portrayed in somewhat human guise. Representations of those deities have often been endowed with the powers of the god as his or her embodiment. One prays before a statue of the Virgin Mary as if praying to the Virgin, or indeed God the Father, her- or himself. Somewhere, at some time, humans have created belief systems in which we can endow the effigy with the presence and powers of the deity.

More often than not, these effigies are powerful as tangible expressions of belief, and it is little wonder that they have been collected as works of art in and of themselves. Why so? Their beauty may be a formal beauty – the use of colour, the skill in carving, the provocative ideas of combining an elephant with a dog, the revelation of the structure of an animal with its external appearance (as in the native carvings of the Northwest Coasts of America and Canada), or even in its basic materials. But in most cases it is not just beauty that is involved. The effigy, the mask and the figurine all condense the power of the deity and, as such, become supra-powerful human creations that transcend the hands and mind of their makers. To augment the power of the image, the maker sublimates or eliminates extraneous detail that might detract from the power of the deity to perform its intended devotional duty. Some masks enfold the salient features of the human face into a smoothly modeled surface of solid and void: sockets become more potent than pupils, softly articulated swellings suggest a mouth, stitches collect as a nose, feathers represent hair. There is no attempt at verisimilitude; instead, an abstraction-based hyperrealism governs. Other masks – for example, from the Congo – use geometry and symmetry to create a more ambiguous visage in which simplification and replication literally multiply the effect of the effigy.[9] Abstractions such as these have provided the foundations for the artifacts produced by many tribal peoples in the past; abstraction continues today – not only among tribal peoples, but also among the arts of the developed world.

Of course, there is an ethical issue involved when removing these objects from the settings and the rituals in which they have been used. Photographs of the objects in their original contexts, and films that portray the rituals in which they were employed, support our understanding of their power as effigies. But even so, divorced from the original belief system, we can never truly understand the power that lies in their condensed forms.

Carved wooden altar-pieces from Germany and northern Europe, executed in a Romanesque style, express the belief of their makers by exaggerating proportions, facial expressions and gestures. The resulting forms heightened the viewer's empathy with the representation. We feel the agony of Christ and the grief of Mary, the sufferings of the common folk as well as of the saints (Figure 10.3). Behind these carvings lie both emotion and simplicity; by eliminating the

10.3
Pietà, St. Mary's Church, Hollola, Finland, fourteenth century

extraneous details, the artist has facilitated the transference of feeling and idea. These figures narrate a story to which we can all relate; we sense their truth and the suffering owing in large part to their frank expression. These are radical works, that is, works that return to the roots of thought and emotion.

Abstraction

Abstraction – literally the 'drawing from', relies on a process of sifting and distillation. To illustrate a process of ever-greater abstraction, in 1925 Theo van Doesburg painted a series of images based on a cow for his Bauhaus book, *Principles of Neo-Plastic Art*. In the first painting, van Doesburg portrayed the animal more or less realistically, if somewhat post-impressionistically.[10] His second panel reduced the cow to geometric shapes, while the last two studies reconfigured those shapes to such a degree that the paintings departed almost completely from any resemblance to the original subject. Van Doesburg's exercise reveals that, in simplification-as-reduction, there is always the danger of losing the idea and feeling as well as the form. Instead, I am arguing for an abstraction that distills and concentrates the power of the original, as might a sauce that has been reduced to intensify its flavour.

In that regard, we might also look at a van Doesburg contemporary: the artist Piet Mondrian. Before reaching the non-figurative gridded panels for which he is known, Mondrian passed through an almost violent expressionism, a stage that produced some of his most emotionally vibrant works. In his painting of a windmill, for example, perspective and colours intensified the impact of the structure to a degree impossible to achieve in a realistic style. One could say that, because painting may never truly capture the reality of any structure, it must compensate by creating its own reality. We can also trace the path by which Mondrian moved from expressionism to non-figuration, in particular in his treatment of the tree as a subject. Over the course of relatively few years, the tree evolved from a structure of quasi-realistic branches to one comprised entirely of vertical and horizontal l ines. These, in turn, led to the even more abstracted paintings executed as plus and minus configurations. Again, the painter here seeks essences, seeking what lies behind the surfeit of detail and form by which we normally know a subject.[11]

Building

Belief also propels the creative making of spaces using restricted means. At the start of the seventeenth century, New Mexico lay at the very edge of the Spanish empire, a distant outpost literally standing in the desert. There was, at first, a hope of finding precious minerals, but that hope soon evaporated, and evangelism and colonisation replaced that original quest. If mineral resources were lacking, the ore of spiritual awakening – the native peoples, termed by the Spanish 'Pueblo' or town-living Indians – was plentiful. To execute their project of conversion, Franciscan missionaries constructed compounds that centred on the church, complemented with living quarters for the priest and monks.[12]

There were problems though. For one, the Franciscans were not trained in building, much less in the fine art of architecture. In addition, other than the missionaries themselves, there were few skilled Spanish craftsmen who knew construction. Added to that were the lack of building stone – the material with which they were most familiar – and the nagging aspiration to construct grand churches that embodied the baroque wonder of the Counter-Reformation, the style already dominating religious construction in central Mexico. A difficult situation indeed. But they had faith. Through coercion and conversion they acquired the labour for construction. For the most part, the native Pueblo Indians had built only in puddled mud construction, with wooden beams and purlins to support a roof of earth.[13] The cells they built were small, far too small for a Catholic church.

Nature instructs us in the economy of means; we say that necessity is the mother of invention. The Spanish missionaries and colonists taught the Pueblo peoples about constructing walls of mud bricks rather than puddling, and they increased the span of the churches by using native labour to drag heavy timbers from the mountain forests almost thirty miles away. But those nagging religious aspirations still troubled the missionaries. Early on, however, the builders realised that, by raising the roof over the sanctuary, they could create a slot that focused light directly on the altar: the 'transverse clerestory' (Figure 10.4).[14] As there were no real windows – perhaps only a few small openings covered with translucent animal skins – the effect of this slot was stupefying.

10.4
Church of San Agustín, Isleta Pueblo, New Mexico. Early seventeenth century; current form, 1960s. The effect of the transverse clerestory

It took some moments for the eye to adjust to the cool darkness of the church. But even while still at minimal aperture, the eye sensed the illumination striking the altar – a glow that increased in brilliance as the eye adjusted to the dimness of the nave. Using literally nothing – that is, nothing except light and soil – the church builders were able to create a sense of wonder especially effective in relation to the place, available materials and the people it served.

Simplicity should not be taken as reduction per se, but instead a husbanding of factors to yield a rich set of experiences over time. Le Corbusier's celebrated Villa Savoye (1931) appears more or less the same from a distance or from close up, in terms of the building's surface textures and details (spatial effect, of course, is another matter). In contrast, the overall form and profile of the refectory at the Cranbrook School for Boys in Bloomfield Hills, Michigan (c. 1925), appear prismatic and symmetrical from a distance (Figure 10.5).[15] But as one approaches this fine work by Eliel Saarinen, one notices that the apertures are *not* symmetrical, and that the play between the dynamic balance of elements within a symmetrical frame energises the composition and provokes interest. As we approach closer still, the varied detailing of the brickwork constantly changes, on first visit and over time. Saarinen, of course, was trying to balance Classical ideals with an Arts and Crafts sensibility, and in meshing the two schools he successfully gained the best from each.

Compare the Cranbrook school with the International Passenger Terminal in Yokohama by Foreign Office Architects, completed in 2002 (Figure 10.6).

10.5
Eliel Saarinen, Cranbrook School for Boys, Bloomfield Hills, Michigan, 1925. Approaching the Refectory

10.6
Foreign Office Architects, International Passenger Terminal, Yokohama, Japan, 2002

The architects' conceit was to be able to walk on the roof of the pier, pulling the earth more deeply into the bay. The computer renderings of the project looked stunning, as computer renderings often do, but the experience of the place is actually not of equal interest. Of course, being able to walk next to the water is attractive and appealing. But the scheme, for some reason, is symmetrical. Why make a 'landscape' symmetrical (unless the demands of structure or the programme of the spaces below so rule)? Here, the idea is simple and powerful as a concept, but the experience is hardly rich – one might glean most of it from the computer rendering or an aerial view, without ever traveling to Yokohama. The slopes of the grass planes are meager, I imagine to avoid ramps and stairs, and the details are prosaic. The wood decking is treated like asphalt, laid from side to side with no sense of edge or transitions, and the design of the railings – of which there are seemingly miles – is banal. Unlike in encountering Saarinen's work, we get little more by joining in the promenade, except to enjoy the effect of the water itself. As a totality, the place feels neither sufficiently dramatic nor sufficiently subtle, which I admit seems like a contradiction.

Time and experience

Some years ago, an American cold medicine called Contac displayed its colourful grains within a clear gelatin capsule. The firm's advertising explained that the different colours indicated the times when the medication would kick in: red was x hours, blue, some time thereafter, and so on – totaling up to twelve hours. It would seem that great architecture and landscapes, at least in terms of our being there, should possess similar properties. We might get the overall form, the global impact, on first take. But there must be sufficient provocation for our wanting to remain there. Over time, new aspects of the place would be revealed, or would be discovered.

The traditional Japanese tea-house provides an excellent example of 'Contac' architecture. Although the tea-house often used the *sukiya* style based on the lowly rural farmhouse, its design was hardly mundane and unselfconscious.[16] To the contrary, every aspect of its design had been careful considered and crafted. The thatch roof was executed in a masterful way, often combining planes of varied slopes in manners that would never have topped a farmhouse (Figure 10.7). The materials and detailing were exquisitely applied and crafted. Inside the building, tranquility reigned, and it was said that all those who entered by crawling through

10.7
Shokin-tei (tea-house), Katsura Villa, Kyoto, Japan, seventeenth century

the small door shared equal social status. The interior was elegantly neutral, the basic orthogonality of the space articulated by the twisting naturalism of the *tokobashira* post. Most of all, the dusky light that softly illuminated the room changed subtly with the passing of the hours; late afternoon was preferred, as the hours slipped from daylight into evening.

This aspect of time was crucial to the execution of the tea ceremony and for the appreciation of the room. Because the space *appeared* so simple, one looked more carefully, perhaps first noticing the beauty of the flower arrangement in the *tokonoma*, the appropriateness of the hanging painting to the current season, the texture of the mud plaster, the glaze of the tea bowl or the obscure light filtered through one or more layers of rice paper. Time creates the sense of event, and architecture informed by time endures in our consciousness. Interestingly, the Japanese character for *ma*, or space, contains elements that suggest both *time* and *place*: the 'gateway' radical for place, that for the 'sun', or 'day', for time.[17]

The nature of light

Trapping light, of course, is nothing new to architects. From the baroque fascination with washes of light that dissolved stuccoed walls and painted images, to the theatrical displays of Alvar Aalto – particularly at the Vuoksenniska church in Imatra – to the skylight and roving beams of Tadao Ando, light has been an important element in the architect's vocabulary. But we still might consider how simply we can construct the instrument for maximal effect.

Light is cheap, although catching and containing it may be less so. By appropriating light from an adjacent space, or through architectural construction, in his installations James Turrell has redirected human attention from the structure of the space to the nature of light, making us conscious of ourselves perceiving.[18] In the 1970s, Turrell purchased an extinct volcano in northern Arizona and, over the course of three decades, he has transformed its cone into a series of artworks (Figure 10.8). Principal among these efforts was the reshaping of the crater into a giant device for experiencing celestial vaulting, the apparent curvature of the heavens. To achieve this effect required the movement of thousands of cubic yards, perhaps hundreds of thousands of cubic yards, of cinder – and the effect is splendid. But, as one who appreciates minimal means, I cite another artwork executed in the early 1970s, about the same time Turrell began work on Roden Crater.

Walter de Maria constructed the *Lightning Field* in 1972 by using hundreds of stainless-steel poles arranged on a grid measuring one kilometre by one mile.[19] The poles are set so that their tops maintain a constant elevation despite the variation in topography, and in that sense they collectively suggest a virtual plane sitting x feet above the terrain. In actual fact, there are relatively few manners in which one can view the *Lightning Field*: those following the stance of classical perspective; those looking at the poles straight on; and those taken from outside the field, flattening perspective into an elevation that lacks all perceived depth. Attention on the poles as objects soon flags, however, and rather than focusing on the work's formal structure, one surveys the terrain and the surrounding mesas – and, most of all, the effects of light. From dawn to noon, to sunset and twilight, and

10.8
James Turrell, Roden Crater, 1979–2007 and continuing. The crater has been reshaped to heighten the experience of celestial vaulting

into evening. Readings of the piece change constantly – an effect heightened by our moving around the poles, repositioning the light source from front to back. Ultimately, the *Lightning Field*, at least as I experienced it, became an instrument for perceiving light, nature, change and the passage of time. The effect was enormous, yet effected using only very minimal means.

Stillness

There is a power to be gained by the focus that accompanies simplicity. For example, the architecture of the Shakers – the common name for the United Society of Believers in Christ's Second Appearing – physically represented the sect's social practices and appreciation of utility.[20] The Shakers honoured the equality of the sexes, although, as a celibate community, men and women slept and worked in separate quarters. This equality with separation led to symmetrical structures with paired entrances. In their meeting houses, several of which were designed in the nineteenth century by Brother Moses Wolson, two entrance doors led to a common room for worship. The interior normally took the form of a column-free room to allow for their dances and chants, which at times reached the level of the ecstatic. There is great beauty in these rooms, with their white walls and their wooden trim, whether painted a chalky blue or left as natural wood.

Mother Ann Lee, the founder of the sect, set its guiding motto as 'hands to work and hearts to God', a credo that led to a cleanliness and functionalism rarely, if ever, matched in the United States either before or after the years in which the Shakers thrived. Their spaces evince an overall simplicity and order, although a more careful regard reveals an attention to detail and craft that produced spaces of continuing interest. Beauty may derive from the interior light warmed by the honey-coloured wooden side panels. Or perhaps it is their intelligence, evident in spanning the spaces of a circular stone barn that resolves the thorny problem of beam spacing between an interior core and an exterior wall. Where practical, they used built-in furniture to secure the visual tranquility that reflected their placid way of life. These cupboards in an attic floor of the settlement at Pleasant Hill, Kentucky, tell us a story of a simplicity rich in its readings, a simplicity where light and wood and detail together condense a system of belief into built form (Figure 10.9).[21]

The architecture of Luis Barragán offers us a sublime example of an architecture of stillness, one in which an apparent simplicity disguises an engagement with space and materials over time. While Barragán's use of strident colour is a hallmark of his work – for example, at the Cuadra San Cristobál outside Mexico City – it is in the configuration of the spaces and their relation to one another that is the true meat of the designs. Often conceived as cellular structures juxtaposed one against another, the rooms of a Barragán house stand somewhat as worlds unto themselves.[22] The house takes the form of a community, although in a manner quite different from, say, that of Aldo van Eyck's orphanage in Amsterdam.[23] One is hard-pressed to draw a plan of a Barragán house after a visit, or even after multiple visits, as the design idea is rooted in the individual unit rather than in any organising scheme, except one planned rather loosely.

In Barragán's own house, one enters and turns on a path that psychologically enlarges the restrictive dimensions of the spaces (Figure 10.10). And then one comes to rest. The materials are usually natural, at times raw, with walls serving as supports for key artworks chosen as much for their architectonic effect as for their surface textures, colours or images. In the small breakfast room, Barragán underlined his quest for meditative tranquility. It is a simple space, looking

10.9
Center Family Dwelling House, Pleasant Hill Shaker Village, Kentucky, 1834. Attic storage

10.10
Luis Barragán, Barragán house, Mexico City, Mexico, 1948. Entry hall

to the garden. The word *soledad* (solitude) on a plate captions his ambitions for this tiny room, enlarged with a view to the garden and with colour.

Lest you assume that only an orthogonal geometry produces simplicity as an expression of authenticity and/or faith, let me offer the example of a small chapel by Eero Saarinen on the MIT campus in Cambridge, Massachusetts.[24] The building takes the form of a cylinder of brick, entered through a short, glass-walled corridor. A screen by Harry Bertoia collects the light falling from the central skylight over the altar and renders it palpable (Figure 10.11). The chapel walls rest on shallow arches that span the surface of a reflecting pool. As the wind blows across the water on the exterior, reflections of light from its surface play in endless variation at the base of the walls, softly enlivening the interior – joining equilibrium with movement and variation.

The geometry of the container is of less concern than the nature of the space contained – an old Taoist idea adopted by Frank Lloyd Wright and other architects in their work. I believe that, today, too many buildings are conceived from the exterior, whether using virtual or physical models, while the nature of the interior spaces is left to chance as a mere by-product. The interior of the passenger terminal at Yokohama, for example, is rather dismal and troubled by glare – all efforts were seemingly placed on the roof-top above. Many strangely configured buildings have

10.11
Eero Saarinen, Kresge Chapel, Massachusetts Institute of Technology, Cambridge, Massachusetts, 1954. Lighting on the altar

interiors that are less than satisfying, a characteristic of many of Frank Gehry's works, for example.[25] Dynamic spaces can result from a far less convoluted frame.

The paragon of rectilinear dynamism is the seventeenth-century Katsura Villa outside Kyoto. The frame of the building is completely orthogonal, using a pre-cut and fitted timber structure infilled with panels of cardboard, rice paper and *tatami*.[26] The module is consistent horizontally and vertically, roughly three feet by six feet. The container, thus, is simple and unassuming, although incredibly sophisticated in its detailing and craft. The villa's three pavilions, its *shoin*, were constructed over a period of nearly a century. Because the Japanese believed that buildings, like natural forms, should evince a sense of completion at every stage of their existence – birth, growth, maturity, decay – the three pavilions were staggered in what is termed a 'geese in flight' arrangement, that is to say, as a loose V. An implied diagonal figure emerges without any true statement of a diagonal line.

Inside the pavilions is a world of dynamic space. The sliding panels, and the fluid spaces that result, banish any vestige of a static architectural world. Equilibrium reigns, as does an overall sense of tranquility. Yet it is an architecture alive, nonetheless. *Shoji* panels modulate congress with the garden, at times occluding certain vistas, on other occasions heightening their effect. It is a spatially dynamic world, although the organising frame itself is stolid and unchanging (see Figure 10.7). This stands in strong contrast to so many of today's buildings, where the shell is highly wrought, but the interior spaces are static and banal.[27]

We may appreciate an overly complex architecture as an image, but I question the effect on us over time. We may be lured, and at first charmed, by the

10.12
Louis Kahn, Salk Institute, La Jolla, California, 1965. Courtyard

complexity of form, although even the seemingly static and simple may speak to us. The buildings at Louis Kahn's Salk Institute in La Jolla, California, stand as two linear blocks, each articulated as stubby towers built with what may possibly be the world's most perfect concrete work. Yet the space between these solid blocks is a void rendered positive, its tiny fountain and aqueous rill challenging the equilibrium of the space and driving us onward (Figure 10.12). Not incidentally, this tiny piece of water links us conceptually, as well, to the great ocean beyond, fusing land and sea.[28]

In closing, allow me to restate my basic premise. If we regard simplicity as condensation or distillation, rather than as reduction and elimination, we may achieve places that are less intrusive, more rewarding, less merely fashionable and more enduring. I believe that, in this process of condensation, we may create an architecture that engages people more profoundly while being more sympathetic to its environment. Simplicity may not always provide the jazzy image so beloved of the student, especially when viewed on a computer screen. But it will be far more substantial in a life rendered, not as a virtual reality, but within the reality in which we live.

Notes

1 I realise that, by using the word 'resist', I may be trampling on Kenneth Frampton's copyright on the term, so *caveat lector*. Kenneth Frampton, 'Towards a Critical Regionalism: Six Points for an Architecture of Resistance', in Hal Foster (ed.), *The Anti-aesthetic: Essays on Postmodern Culture*, Seattle, WA: Bay View Press, 1983, pp. 16–30. I have examined the idea of critical regionalism applied to landscape design in 'Aspects of Regionality and the Modern(ist) Farden in California',

in Therese O'Malley and Marc Treib (eds), *Regional Garden Design in the United States*, Washington, DC: Dumbarton Oaks, 1995, pp. 5–42.

2 'Nothing is so hard as to preserve a silence in a tumultuous urban life.' Takefumi Aida, 'Silence', in Kenneth Frampton (ed.), *A New Wave of Japanese Architecture*, New York: Institute for Architecture and Urban Studies, 1978, p. 14. See also Takefumi Aida, *Takefumi Aida: Buildings and Projects*, New York: Princeton Architectural Press, 1990, although the book centres on later work.

3 Although Ando's mature work has continued in the restraint of his early work, the scale of projects has demanded a corresponding increase in complexity. For a survey of his architecture, see Francesco Dal Co, *Tadao Ando*, London: Phaidon, 1995, especially pp. 34–65; and Philip Jodidio, *Tadao Ando: Complete Works*, Cologne: Taschen, 2004, pp. 38–47.

4 Although Aarno Ruusuvuori produced only a limited number of works, each of them was exquisitely detailed and executed. Among his finest works was the church in Tapiola, completed in 1965, which had a considerable influence on the young Tadao Ando. See Marja-Riitta Norri and Maija Kärkkäinen, *Aarno Ruusuvuori: Structure is the Key to Beauty*, Helsinki: Museum of Finnish Architecture, 1992; and Kirsi Leiman, *Concrete Spaces: Architect Aarno Ruusuvuori's Works from the 1960s*, Helsinki: Museum of Finnish Architecture, 2000.

5 Darcy Wentworth Thompson, *On Growth and Form* (1917), reprint, New York: Cambridge University Press, 1961.

6 David Hancock, *Master Builders of the Animal World*, New York: Harper & Row, 1973, pp. 129–37.

7 See Martin Filler, 'Santiago Calatrava', in *Makers of Modern Architecture: From Frank Lloyd Wright to Frank Gehry*, New York: New York Review of Books, 2007, pp. 283–98. My disaffection with Calatrava's work is architectural rather than structural, however: for example, in its streamlining, or its lack of a constellation of details revealed over time. I prefer my engineering less stylised.

8 See Sokyo Ono, *Shinto: The Kami Way*, Rutland, VT: Charles E. Tuttle, 1962, and C. Scott Littleton, *Understanding Shinto*, London: Duncan Baird, 2002.

9 The references here, although more broadly applicable, are specifically derived from masks in the collection of the Barbier-Müller Museum in Geneva, Switzerland.

10 Theo van Doesburg, *Principles of Neo-Plastic Art* (1925), trans. Janet Seligman, New York: New York Graphic Society, 1966.

11 Publications on Mondrian are numerous. Among the most comprehensive are Angelica Rudenstine (ed.), *Piet Mondrian, 1872–1944*, Washington, DC: National Gallery of Art *et al.*, 1994; and Carol Blotkamp, *Piet Mondrian: The Art of Destruction*, New York: Harry Abrams, 1994; for Mondrian's own thoughts, in his own words, see Piet Mondrian, Harry Holtzman, and Martin S. James (eds and translators), *The New Art – The New Life: The Collected Writings of Piet Mondrian* (1986), reprint, New York: Da Capo Press, 1993.

12 See Marc Treib, *Sanctuaries of Spanish New Mexico*, Berkeley, CA: University of California Press, 1993.

13 On the Pueblos, see W. C. Sturtevant and O. Alfonso (eds), *Southwest, Handbook of North American Indians*, Vol. 9, Washington, DC: Smithsonian Institution, 1979, and Vincent Scully, *Pueblo: Mountain, Village, Dance*, New York: Viking Press, 1972.

14 See Treib, *Sanctuaries*, pp. 36; 41–3; George Kubler, *The Religious Architecture of New Mexico: In the Colonial Period and Since the American Occupation*, Albuquerque, NM: University of New Mexico Press, 1990, pp. 66–7.

15 Saarinen came to Cranbrook from his native Finland in 1923, after his success in the Chicago Tribune Tower competition. From 1924, until nearly his death in 1950, he designed buildings for the Cranbrook complex: the Boys' School, the Kingswood School for Girls, and the Cranbrook Academy of Art. See Adele Westbrook and Anne Yarowsky (eds), *Design in America: The Cranbrook Vision, 1925–1950*, New York: Harry Abrams, 1983; and Gregory Wittkopp (ed.), *Saarinen House and Garden: A Total Work of Art*, New York: Harry Abrams, 1995.

16 The classic discussion of the tea ceremony and its architecture is Kakuzo Okakura, *The Book of Tea* (1906), reprint, Rutland, VT: Charles E. Tuttle, 1956. See also Tatsusaburo Hayashiya, Masao Nakamura and Seizo Hayashiya, *Japanese Arts and the Tea Ceremony*, trans. Joseph P. Macadam, New York: Weatherhill, 1974.

17 Arata Isozaki, *Ma: Space-Time in Japan*, New York: Cooper-Hewitt Museum, 1979.

18 Among the most comprehensive presentations of Turrell's works are: Julia Brown (ed.), *Occluded Front: James Turrell*, Los Angeles, CA: Fellows of Contemporary Art and the Museum of

Contemporary Art, and Larkspur Landing: The Lapis Press, 1985; Lynn M. Herbert *et al.*, *James Turrell: Spirit and Light*, Houston, TX: Contemporary Arts Museum, 1998; Peter Noever (ed.), *James Turrell: The Other Horizon*, Ostfildern-Ruit: Cantz Verlag, 1999; and Craig Adcock, *James Turrell: The Art of Light and Space*, Berkeley, CA: University of California Press, 1990.

19 'The invisible is real.' 'Isolation is the essence of Land Art.' These two quotes are taken from an article by the artist presenting the *Lightning Field*: Walter de Maria, 'Some Facts, Notes, Data, Information, Statistics and Statements', *Artforum*, April 1980, p. 58. See also John Beardsley, *Earthworks and Beyond: Contemporary Art in the Landscape*, New York: Abbeville Press, 1984, pp. 59–63.

20 See Edward Deming Andrews, *The People Called Shakers*, New York: Dover Publications, 1953; and Marguerite Fellows Melcher, *The Shaker Adventure*, Cleveland, OH: The Press of the Case Western Reserve University, 1968.

21 Thomas D. Clark and F. Gerald Ham, *Pleasant Hill and Its Shakers*, Harrodsburg, KY: Pleasant Hill Press, 1987.

22 The plan of the Barragán house is beautifully presented in Wim van der Bergh, *Luis Barragán: The Eye Embodied*, Maastricht: Pale Pink Publishers, 2006. On Barragán's architecture in general, see also Federica Zanco (ed.), *Luis Barragán: The Quiet Revolution*, Milan: Skira, 2001; Antonio Riggen Martínez, *Luis Barragán: Mexico's Modern Master, 1902–1988*, New York: Monacelli Press, 1996; and Raúl Rispa (ed.), *Barragán: The Complete Works*, New York: Princeton Architectural Press, 1996.

23 Completed in 1961, the orphanage used an agglomeration of several types of module based on the square. The effect is far more regular in its variation than Barragán's more personal application of variation, based on emotional and functional necessity. See Francis Strauven, *Aldo van Eyck: The Shape of Relativity*, Amsterdam: Architectura & Natura, 1998, pp. 284–324.

24 On Eero Saarinen and the Kresge Chapel, see Aline B. Saarinen, *Eero Saarinen on His Work*, New Haven, CT: Yale University Press, 1962, especially pp. 40–5.

> I have always remembered one night on my travels as a student when I sat in a mountain village in Sparta. There was bright moonlight over head and then there was a soft, hushed secondary light around the horizon. That sort of bilateral lighting seemed best to achieve this other-worldly sense. Thus, the central light would come from above the altar – dramatised by the shimmering golden screen by Harry Bertoia – and the secondary light would be reflected up from the surrounding moat through the arches. (p. 42)

See also Eeva-Liisa Pelkonen and Donald Albrecht, *Eero Saarinen: Shaping the Future*, New Haven, CT: Yale University Press, 2006.

25 A notable exception, however, is Gehry's Peter Lewis Building at Case Western Reserve University in Cleveland, Ohio, where the lobby spaces assume the vibrant shapes of eroded canyons from the American Southwest. These spaces flow in an intriguing way, luring us on. It is also a tour de force of construction technique, managing to extract a double curvature from flat sheet-rock panels.

26 On Katsura, see Akira Naito and Takeshi Nishikawa, *Katsura: A Princely Retreat*, Tokyo: Kodansha, 1994; Virginia Ponciroli (ed.), *Katsura: Imperial Villa*, Milan: Electa Architecture, 2005; and the classic Kenzo Tange, *Katsura: Tradition and Creation in Japanese Architecture*, New Haven, CT: Yale University Press, 1960.

27 I will avoid providing further examples of that misfortune, however.

28 Marc Treib, 'To End a Continent: The Courtyard of the Salk Institute', *Journal of the Society of Architectural Historians*, September 2006, pp. 402–27.

Chapter 11

'Stirling's worth'

The quality of the Florey Building, Oxford

Igea Troiani

In 1976, Gavin Stamp published a critical review of one of James Stirling's most famous buildings. Questioning provocatively the value of the architect and his building, the text is titled ' "Stirling's Worth": The History Faculty Building'.[1] Written from the perspective of an unhappy user of the building, Stamp challenges the assessment, by architecture critic Reyner Banham, of the quality and performance of what is recognized by many architectural connoisseurs as one of Stirling's best buildings, designed in his 'experimental modernist' phase, 1950–74.[2] Apart from the History Faculty Building at the University of Cambridge, 1964–7, such early period works also include Stirling and Gowan's Engineering Building at the University of Leicester, 1959–63; and the lesser known Florey Building at The Queen's College in Oxford, 1966–71 (Figure 11.1).

In a similar vein, this paper discusses Stirling's architectural worth in relation to the Florey Building. In an effort to reveal the variance in any definition of architectural quality, it highlights the schism between the client and the architectural establishment's view at the time. Through a reading of archival material held at the Canadian Centre for Architecture (CCA) in Montreal and The Queen's College in Oxford, the paper reveals the reasons why the commissioning client group and the Florey Building's users sought to undermine the value of Stirling's architecture. This will be contrasted to the reasons why architectural connoisseurs, both nationally and internationally, praised the building for its innovation, heralding Stirling as a new figurehead for the profession.

11.1
Photograph of model of Florey Building, in The Queen's College Archive, dated 1968. Photographer unknown

The Florey Building

The idea to construct a new student accommodation building at The Queen's College came from Lord Howard Florey, Provost from 1962 to 1968 and Nobel Prize winner in 1945. Built in his honour, Florey was one of the Building Committee members who, on 4 August 1964, chose the Glaswegian-born architect James Stirling (1924–92) to design the building at a site overlooking the River Cherwell (Figure 11.2). Others on the committee were Professor of History, John Prestwich; the Senior Tutor and French Don, Professor Iain Macdonald; and Bursar, Arthur Williams. Macdonald fully supported appointing Stirling. According to Prestwich, Macdonald 'was a great believer in the virtues of all Scots, particularly Glaswegians'.[3] Prestwich favoured Ahrends, Burton, Koralek (ABK). Both Florey and Prestwich were strong supporters of modern architectural design. They felt that Stirling, although not the architect favoured by consensus, could offer them a signature building that would attract students on the basis of its modernity. The committee were aware that Stirling was gaining a national reputation. Arguably, they might also have been impressed by how highly Stirling was regarded internationally, particularly in America, where, since 1960, he had been a visiting teacher at the School of Architecture at Yale University. Williams, however, was opposed to Stirling's appointment. In *Big Jim: The Life and Work of James Stirling*, Mark Girouard describes Williams as having been 'against any new building at all', because from early financial predictions it would be unfeasible, and adds, 'he and Jim never got on'.[4]

11.2
View from the River Cherwell of the Florey Building, Queen's College, Oxford, 1972

The client's perspective

The most concise chronology of the Florey Building's construction history, as seen from the client's perspective, was prepared by Williams in 1981, ten years after the building's completion. It is a twenty-nine-page report titled *The Florey Building: A Narrative Account of the Project.*[5] The report summarizes the content of six archival files of project correspondence held at The Queen's College.[6]

The Florey Building took much longer to build than promised: seventeen months more than the original twenty-one-month building period. The construction period was in part extended because Stirling's site architect for the project, Roy Cameron, was young and not permitted to make on-site decisions. At the time of construction, Stirling was flying between the United States and Britain, working on projects and teaching in both countries. Resolution of site queries was therefore a protracted process. Also contributing to the claims for extensions of time was the recurrent delay by Stirling's office to provide detailed architectural drawings in advance to the contractor, W.H. Chivers & Sons. This deferred and sometimes halted site work.

These extensions of time were also the consequence of problems with Stirling's tiling and glazing specifications for the building.[7] For the Florey Building, the same specification for adhering red tiles to façades and courtyards was specified as had earlier been used at Leicester and Cambridge. Like other tile-clad buildings built in this period, façade tile repair is an ongoing issue. Glazing on the Florey Building was inadequately detailed by Stirling's office, such that different component parts had not been included in the original contract documents. The college was then made aware that Stirling had omitted ventilation louvres, a major issue with such a vast glass façade on the interior of this amphitheatre-shaped building.

Nervous about rising costs, Williams made the college aware that the situation might become litigious.[8] Obviously, Stirling was also conscious of his risk. He approached the college, asking that an RIBA agreement be signed between the two parties. The college declined, since the building construction had then been in progress for almost one and a half years. Williams writes:

> His [Stirling's] signal failure to reply to requests for written reports on the causes of increase of cost and to account for the numerous variations and additions which he was authorising [. . .] led to the Bursar consulting other authorities who were known to be dissatisfied with his work for them, being the University of Cambridge in respect of its new History Faculty Library and the University of Leicester in respect to its Engineering Laboratory. From these sources it was learned that a firm of solicitors in London had come to specialise in the grievances of various of Stirling's clients, and they were engaged to advice [*sic*] the College on its remedies.[9]

On the advice of the solicitors Walters Vandercom & Hart, an independent architectural arbitrator, Leslie Alexander, was commissioned to review Stirling's professional performance. In a meeting titled 'Without prejudice', Alexander raises the issue of the consequences of a client appointing an avant-garde architect. Importantly, he notes: 'By going to Stirling a Client anticipated a very modern building – but one designed with adequate and proper skill.'[10]

Williams's narrative account of the project concludes by reflecting pessimistically on 'The Achievement'. In it, he refers to Stamp's review of Stirling's Cambridge building. For Williams, Stamp's criticisms of 'Stirling's worth' are as valid in the context of the Oxford building as the Cambridge one. Aside from the arduous building process experienced, the final building does not perform well functionally. Williams argues the practical problems with their buildings result from Stirling's attitude to architecture. He delivers on the avant-garde aesthetic and the moral conviction to architecture, but do the buildings work?

Stamp describes a Cambridge building that – owing to the way it was designed, with large expanses of glazing – suffers from extremes of heat or cold, has glare and ventilation problems, and is difficult to clean and main faults hold true for the Florey Building. But, in addition, at the Florey Building are issues with privacy, a consequence of its plan form. According to Geoffrey Tyack, the Florey Building's u-shaped plan means that students 'are obliged to contemplate each other in their box-like split-level rooms like goldfish in a bowl' (Figure 11.3).[11]

The Cambridge History Faculty leaks, water dripping through ceilings and down walls. Leaking has also been a problem at the Florey Building. Water penetration has meant that tiling of ground areas and on wall façades needed remedial work. The proximity of the Florey Building to the river might explain the issues with subsidence, a factor contributing to the tiling problem. In the Stirling Archive held at the CCA, there are hundreds of photographs showing water defects at the History Faculty building. Like the Cambridge users who, Stamp reveals, did not want to inhabit the History Faculty building when first completed, Williams explains: 'The first generation of occupants of the Florey Building had in fact presented a petition to the College asking to be allowed to live elsewhere'.[12]

11.3
Glazing façade of the Florey Building, 1972

As soon as the Florey Building was completed and the contractor paid out, the College issued Stirling a writ, which he was astonished to receive. All of the building problems had contributed to the escalation of project cost from £263,865 to £355,299. 'This was an excess of £91,434 [. . .] and of this [. . .] only £37,376 had been agreed by the College as the cost of extras which it had authorised'.[13] Of the three university clients, Queen's College was the only client successful in its litigation, Stirling's insurers offering £25,000 compensation.

As a consequence of his 'disastrous'[14] Florey experience, to quote Williams, and knowing the difficulties the University of Leicester and the University of Cambridge had experienced, Williams and the new Provost, Lord Blake, began to rally other university clients against appointing Stirling. In 1972, Blake wrote to the Registrar at the university to state his opposition to: 'any suggestion that he (Stirling) should have any part in any new university building'.[15]

The architectural connoisseur's perspective

Mark Girouard wrote:

> The Engineering Building at Leicester made him [Stirling] internationally famous. It and its successors were given top billing in architectural magazines everywhere [. . .] Le Corbusier was getting old and his star was on the wane. His last buildings received comparatively little attention, and he died in 1965. Architects and, perhaps even more, architectural students, were looking for a new god, and in Jim, more than in any other architect of his generation, they found one.[16]

Stirling's high status in the architectural profession was established immediately from his and Gowan's Leicester Engineering building. Arguably filling the role of Le Corbusier, Stirling's early career work for universities was taken up by the architectural press, and Stirling – not Gowan – was hailed 'a new god'. Starting in 1962, there was a steady stream of publicity about the university projects in the popular architectural media, which fuelled architectural practitioners', critics' and students' fascination with Stirling. Many flocked to see and experience the building. Some architects moved to be taught by, or work with, their guru. Girouard claims that Craig Hodgetts was one so inspired: he left his wife and moved to study at Yale University so as to be instructed by Stirling. Another was Leon Krier, who moved from Stuttgart to London to work in Stirling's office after seeing the Engineering building, because 'it seemed like something fantastic'.[17]

But not only practitioners and students were infatuated with this early architecture in the late 1960s and early 1970s; numerous architectural critics extolled the virtues of the Cambridge and Oxford buildings. The impressive list includes, among others, Alvin Boyarsky, Reyner Banham, Arthur Drexler, Charles Jencks and Joseph Rykwert. In the essay 'Stirling "Dimostrationi" ', published in the October 1968 issue of *Architectural Design*, Boyarsky elects to construct a picture of the Cambridge History Faculty as exemplary of Stirling's demonstration against the then British direction of modern architecture advocated by German émigré Nikolaus Pevsner. Described literally to be in battle, Boyarsky presents

Stirling as fighting 'by absolute and uncompromising moral views; a guerrilla like undermining of establishments and traditions'.[18] Soon after, the Cambridge History Faculty is reviewed by Banham in the *Architectural Review*. In his article, Banham focuses on the issue of the building's technologies. He praises Stirling's tile and glass formula, which results in a successfully varied work in the Engineering building 'style'. In the vein of his *Architecture of the Well-Tempered Environment*, Banham reviews the building's thermal performance, arguing that, on the summer's day he visited, he did not need to remove his jacket and, as a consequence, gives the building a nod of approval for its environmental comfort.[19]

Having built up media momentum, in the same year the Florey Building is presented to the American public, even though its construction had barely begun. Alongside the completed Leicester Engineering building and partially completed Cambridge History Faculty, it is exhibited through models and drawings in the Museum of Modern Art (MOMA) New York exhibition, *James Stirling: Three University Buildings*.[20] In a press release, Drexler, Director of the Museum's Department of Architecture and Design and exhibition curator, pays tribute to the buildings as model examples of a new architecture. He commends them for their 'vitality' and refreshing 'complexity', but in particular for 'their *quality* of principled and rational purpose, sustained by a subtle sense of form and material'.[21]

Jencks discusses the three university buildings in *Modern Movements in Architecture*. In it, he separates out a discussion of the Florey Building. As if describing the other two, he admires its 'shimmering prismatic glass versus bright red tile' (Figure 11.4).[22] He commends Stirling's rational approach and claims: 'For the first time since the Palm House at Kew or the Crystal Palace, Britain had a designer who could handle glass with virility.'[23] The virile handling of glass versus red brick and tile elements is compared to the work of the Russian constructivists. By architectural association or not, in Jencks's view, Stirling is 'the best architect of his generation'.[24] In the November 1972 issue of *Domus*, Rykwert also focuses on the composition of Stirling's university architecture. Rykwert presents the argument that, of the three university buildings, the Florey is 'Stirling's most mannerist work'.[25] Although the simplest of the three, Rykwert deems it the most carefully composed.

Soon after Rykwert's review was published, Stirling was sued by The Queen's College. From then on, the Florey building received virtually no architectural publicity. It almost vanished from the architect's *oeuvre* and the radar of the architectural press. After the Florey building, Stirling's architectural 'style' changed markedly. His aesthetic, reminiscent of the constructivists, disappeared. He abandoned the red brick tile and glass formula. No longer did he feel the commitment – which John Kean contends first appeared even earlier than the Leicester Engineering building – to use tiles laid vertically as cladding in order to accentuate the surface, rather than structure, of the separate building forms.[26] Stirling's client base changed: he won the majority of his commissions from Germany and the United States and had no commissions in Britain for almost ten years. Stirling and his followers blamed the British for not appreciating the modern look of his architecture.

11.4
The tiled courtyard and podium of the Florey Building, 1972

That same year, Stirling was awarded the RIBA Royal Gold Medal for Architecture. In 1981, Stirling was awarded the Pritzker Architecture Prize. That was the year Williams wrote his condemnatory report of Stirling's performance on the Florey Building. The jurors on the Pritzker selection committee who chose Stirling were J. Carter Brown, Chairman of the Jury; Lord Clark of Saltwood; J. Irwin Miller; and the eminent architects, Cesar Pelli, Arata Isozaki and Philip Johnson.[27] In relation to the quality of Stirling's early university architecture, the Leicester Engineering building and Cambridge History Faculty building are shown on the Pritzker website, but the Florey Building is not. The jurors' citation is important. It reads:

> In three countries – England, Germany, and the United States – he is influencing the development of architecture through the *quality* of his work'.[28]

Definitions of architectural quality

Stirling is not uncommon in being heralded a great architect by the architectural profession, while knowingly having designed buildings – to which his recognition is attributed – that are flawed in their functional, structural or environmental performance.

Consider the work of the more prolific architect, Frank Lloyd Wright. Amanda Cooper notes, in 'Frank Lloyd Wright's Star Eclipses his Buildings', that, 'while he is lauded as a creative genius, perennial cracking and leaking seem to be a feature of much of his work'.[29] Cooper cites Wright's New York Guggenheim Museum (1956–9) as having had problems with cracking since opening; and one of the all-time best works of American modern architecture, the Fallingwater house (1934), 'has undergone years of work to fix sagging terraces, leaking and cracking'. Sagging, leaking and cracking are tolerated by architectural connoisseurs of Wright's work on the basis that artistic innovation is the paramount criterion of architectural excellence. Some admirers of Wright's work argue that, when being brave and using new technologies to advance architectural design, problems will inevitably arise.

The same arguments are offered to counter detractors of Stirling's early university architecture. As Girouard notes: 'When the [Stirling's] buildings had faults the people who were delighted did their best to extenuate or ignore them.'[30] Banham's aforementioned Cambridge History Faculty review falls tidily in this category. Aware of the client's problems, he suggests that discomfort would more likely be due to the client's lack of expertise in operating the roof's mechanical ventilation system rather than design negligence on the part of the architect.

Stirling's worth in Oxford

Twenty-one years after his first critique of the Cambridge History Faculty building, Stamp argues sarcastically in 'The Durability of Reputation' that the unsuccessful attempt by the University of Cambridge to demolish its History Faculty in 1984 was a good thing.[31] Stamp contends that it is 'important that James Stirling's Cambridge History Faculty should survive in material form to demonstrate just how coarse and thoughtless, how arrogant and inhumane a masterpiece by a world-famous, award winning architect can actually be'.[32]

Stirling's worth has also been questioned, more privately and litigiously, by Williams and The Queen's College, Oxford. After its involvement in a difficult, prolonged construction process, litigation and a thirty-six-year maintenance programme that has not yet obliterated 'teething troubles', The Queen's College remains unsure of the quality of its Florey building.

Unlike Stirling's client critics, architectural connoisseurs agree almost unanimously on Stirling's worth as an architect. Commentators such as Boyarsky, Banham, Drexler, Jencks and Rykwert applaud the experimental modernism Stirling produced between 1950 and 1974. In this body of work, the Florey Building is valued for its bold originality. But the value of Stirling's early architecture lies even deeper. Stirling's early university works offered a spark of hope for a new direction in British architecture, refreshing and divergent from the modernism promoted by

Pevsner. It was daring and anti-establishment. The Florey Building symbolizes a professional victory for architects, a triumph of modern form and aesthetic over tradition. The Florey Building's cultural importance to a profession ideologically disillusioned, then and now, is indisputable. Its quality, however, remains a matter of debate.

Notes

1 Gavin Stamp, '"Stirling's Worth": The History Faculty Building', *Cambridge Review*, 30 January 1976, pp. 77–82.

2 The phrase 'experimental modernism' is drawn from Geoffrey Tyack, *Oxford: An Architectural Guide*, Oxford: Oxford University Press, 1998, p. 323. Tyack notes: 'The Florey Building represents the culmination of Oxford's flirtation with experimental modernism.' Stirling's 'experimental modernist' phase is regarded to be the period documented in James Stirling, *James Stirling: Buildings & Projects 1950–1974*, London: Thames & Hudson, 1996.

3 John Prestwich, quoted in Mark Girouard, *Big Jim: The Life and Work of James Stirling*, London: Chatto & Windus, 1998, p. 157.

4 Mark Girouard, *Big Jim: The Life and Work of James Stirling*, London: Chatto & Windus, 1998, p. 157.

5 Arthur Williams, *The Florey Building: A Narrative Account of the Project*, 28 February 1981, The Queen's College Archive, File no.: 2W.135, pp. 1–29.

6 The six files held at The Queen's College Archive are File nos. FB1788, 'Architect: General File'; FB1796, 'Damage (Various) and Insurance'; FB1797, 'Post-completion Maintenance: Damage and Defects etc.'; FB1798, 'Design: (Inherent Defects of)'; and FB1803 and FB1804, 'Case Against Architect': Box 1 and Box 2, respectively.

7 Electing to discuss these more severe problems of tiling and glazing does not exclude the fact that there were also problems with sound insulation; cracking in ceilings; window staining; and corrosion. All are outlined in the report.

8 Refer to Section 'V. Litigation, 33. Action against Architect' in Arthur Williams, *The Florey Building: A Narrative Account of the Project*, 28 February 1981, The Queen's College Archive, File no.: 2W.135, pp. 26–7.

9 Arthur Williams, *The Florey Building: A Narrative Account of the Project*, 28 February 1981, The Queen's College Archive, File no.: 2W.135, p. 26. Refer to letter to Sir Fraser Noble, then Vice Chancellor at the University of Leicester, from A.A. Williams, dated 26 August 1971, The Queen's College Archive, File no.: FB1788; and letter to Professor W.O. Chadwick, then Master of Selwyn College at the University of Cambridge, from A.A. Williams, dated 26 August 1971, The Queen's College Archive, File no.: FB1788.

10 Leslie Alexander, from 'Record Notes of "Without Prejudice" Meeting held on 10 July 1975 at the Offices of Messrs Masons, London', The Queen's College Archive, File no.: FB1788, 'Architect: General File', p. 1.

11 Geoffrey Tyack, *Oxford: An Architectural Guide*, Oxford: Oxford University Press, 1998, p. 323.

12 Williams, *The Florey Building: A Narrative Account of the Project*, p. 29.

13 Ibid., p. 24.

14 Ibid., p. 29.

15 Letter by Provost Lord Blake, FBA, JP, to Folliot, The Registrar, The University Registry, Clarendon Building, Broad Street, Oxford, dated 26 June 1972.

16 Girouard, *Big Jim*, p. 92.

17 Leon Krier, in Girouard, *Big Jim*, p. 149.

18 Alvin Boyarsky, 'Stirling "Dimostrationi"', *Architectural Design*, October 1968, 38, p. 455.

19 Reyner Banham, 'History Faculty, Cambridge', *Architectural Review*, December 1968, 144, p. 329.

20 The exhibition ran from 17 June to 4 August 1968.

21 Arthur Drexler in 'Press Release No. 53, Museum of Modern Art, New York dated 17 June 1968'; The Queen's College Archive, File no.: FB1788. Italics added.

22 Charles Jencks, *Modern Movements in Architecture*, London: Penguin Books, 1982, p. 269.

23 Ibid., p. 261.

24 Ibid., p. 260.
25 Joseph Rykwert, 'Florey Building, Queen's College, Oxford', *Domus*, November 1972, 516, p. 15.
26 John McKean, *Leicester University Engineering Building: James Stirling and James Gowan*, London: Phaidon, 1994, p. 36.
27 Martha Thorne and Colin Amery (eds), *The Pritzker Architecture Prize: The First Twenty Years*, New York: Harry N. Abrams, Inc., Publishers in association with The Art Institute of Chicago, 1999, p. 179.
28 1981 Pritzker Jury, 'Citation from the Pritzker Jury'. Available at www.pritzkerprize.com/stirling (accessed 20 June 2007). Italics added.
29 Amanda Cooper, 'Frank Lloyd Wright's Star Eclipses His Buildings', *Scotsman*, 16 October 2006. Available at http://news.scotsman.com/ (accessed 14 June 2007).
30 Girouard, *Big Jim*, p. 149.
31 Charles Knevitt, 'Prize University Building To Be Demolished', *The Times*, 19 June 1984.
32 Gavin Stamp, 'The Durability of Reputation: The Cambridge History Faculty', *Harvard Design Magazine*, Fall 1997, p. 57.

Part 4

Measure

In Chapter 12, which begins and ends with Nicholas Grimshaw's new spa building at Bath in the UK, Juliet Odgers refers to an incident in the Cornish town of Camelford in 1988 where, owing to mismanagement at a treatment site, drinking water was poisoned. Although the water tasted repellent, people still consumed it and became very ill because they relied on the authorities' misguided advice that it was drinkable. Odgers suggests that people's capacity to perceive depends on their situation within a culture that supports the acquisition of certain perceptive skills. In recent generations, systematic quality control has, in many spheres, including that of water quality, supplanted a culture of perception and judgement. Is this, Odgers asks, what is also happening in architecture? Quality control attempts to bypass reliance on widespread educated perception. She raises the question of how quality is measured, what criteria are used, and who determines quality on behalf of whom. This fourth and final part of the book explores questions of measure.

The psychologist Nicholas Humphrey has suggested that *qualia*, the qualities of sensory experiences that, taken together, come to determine our identities, emerge in the process of translation from sensation to perception. In Chapter 13, Hugh Campbell relates Humphrey's thinking to the photographs of John Szarowski and the autobiography of architect Louis Sullivan. Qualia, Campbell suggests, emerge autobiographically: 'we become ourselves only in so far as we develop the capacity to differentiate, define and describe that self.' That ability to differentiate – to trust our judgement enough to refuse poisoned water, for example – defines our capacity to measure the world.

In art, the signature is a mark of an object's authenticity, confirming its authorship and completeness, as Paul Emmons reflects in Chapter 14. The signature becomes a measure of quality, an affirmation of status. Signatures recur in Western consumerism, from perfume bottles to shoes to grilling machines, and architects through history have also signed their buildings. Emmons reviews some signed buildings and their manner of signature. Signatures physically inscribe architects' authorship, their feeling for the importance of their own autobiography, into the building fabric, potentially revealing quality as character. The signed building, he argues, is a long way from the signature building.

According to psychoanalyst Jacques Lacan, the earliest architecture was organised around emptiness. Drawing from Lacan, in Chapter 15, Catherine Belsey reflects that what gives most pleasure in cultural artefacts is not subject to the kind of measurement preferred by quality managers. Reflecting – among others – on the most ancient Maltese architecture, less ancient Athenians, Freud, Brunelleschi, *Gone with the Wind*, Titian and *The Third Man*, Belsey argues for the lost object of conscious desire. She finds common ground in the unresolved element. Quality withholds, she suggests, tantalising and seducing, not allowing you to finish with something. It persists where there are no guarantees, in missing elements, and as such is no easy measure.

Chapter 12

Water and the historicity of perception

The New Royal Baths at Bath

Juliet Odgers

This paper is about water: how we perceive it; how we use it; and how our perceptive capacities are shaped by our cultural practices in relation to it. It is also about the New Royal Spa building at Bath by Nicholas Grimshaw and Partners. The Spa facility, which is operated as 'Thermae Bath Spa', is situated right in the centre of the city of Bath (Figure 12.1). In addition to the new building, it includes two refurbished late-eighteenth-century bath houses – the Hot Bath, which is attached to Grimshaw's new structure, and the Cross Bath, which stands alone and has separate street access.[1] As the spa operator's website proclaims: 'Thermae Bath Spa is the only Spa in Great Britain to offer bathing in naturally warm spring waters.'[2]

There is, as the city's name indicates, a long tradition of 'bathing' associated with Bath's thermal springs. From the Roman period through the Middle Ages, right up to the eighteenth and into the nineteenth century, Bath was an important centre of healing practice. In the seventeenth century, before the city's period of expansion fuelled by a fast-growing culture of leisure tourism, Bath was principally known as a place for water cures and was rivalled in England only by London for the number and expertise of its medical practitioners (Figure 12.2).[3] Today, high-end medical practice has shifted to the Royal United Hospital Bath NHS (National Health Service) Trust, an encampment about equal in size to the medieval

(a)

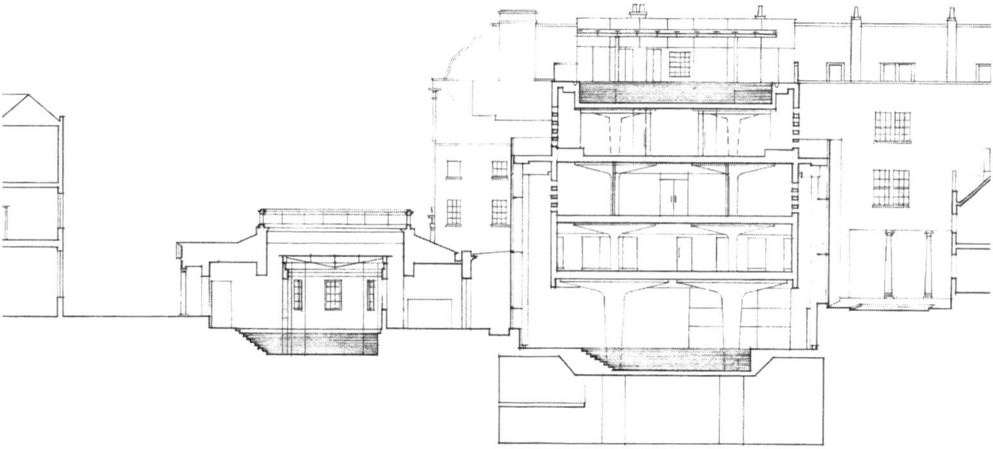

(b)

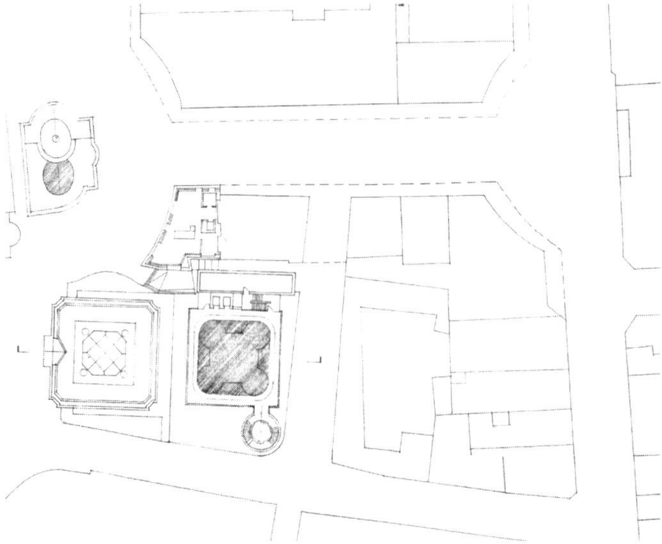

12.1
(a) Section through the New Royal Baths. The Minerva Pool at ground level communicates with the eighteenth-century Hot Bath. Changing facilities, therapy rooms, steam rooms and roof-top bath are stacked one above the other within the stone cube. This in turn is enclosed in a skin of glass.
(b) The New Royal Baths communicates with the Hot Bath to the west. The free-standing Cross Bath, where the thermal spring rises, is to the north west of the new structure

12.2
The King's and Queen's Baths, 1730s – sick and well bathe together (Robins, engraved by Elliot)

heart of the city and situated a few miles outside town. The sick people who used to bathe in the thermal waters have been taken away, and the town and the baths are relinquished to residents and tourists. Medicinal bathing has long since disappeared from the mainstream practices of medicine. How then do we view the experience of 'bathing in naturally warm spring waters offered by the Spa',[4] and how does Grimshaw's design for the New Royal Baths frame this experience?

I will approach this question indirectly through consideration of the lost culture of perception and expertise represented in the writings of two seventeenth-century protagonists – a spa water doctor, John Floyer (1649–1734), and a tourist and spa enthusiast, Celia Fiennes (1662–1741).

Doctor

In 1697, John Floyer enters the annals of Bath's medical history with a pamphlet, *An Enquiry into the Right Use and Abuses of the Hot, Cold, and Temperate Baths in England*, in which he supports the use of tepid and cold baths such as those at Buxton.[5] The Bath doctors did not particularly welcome this contribution, since theirs was a hot spa. A controversy ensued, with Dr Thomas Guidott writing in defence of hot-water bathing. What I wish to draw out from Floyer's writing is his attitude to qualities, perception and instrumentation. Quite typically for his time, he is caught up in a dialogue of traditional practice and new experimental philosophy. His conceptual framework of the healthy body held in harmonic balance with the cosmos is still, to a large extent, classical. It relies on ideas about balancing the humoural qualities of the body as a way of attaining and preserving health, thus:

> That all these kind of Bathings are very useful, the Experience of all Ages testifies; and they always applied them to different States of our Humours; For if we be too hot our Reason, as well as Experience, prescribes cold Bathing in *Ephemera's* and excessive Heats; But if our Bodies be chill, and pained, we use hot *Baths;* and for Pleasure, cleansing the Skin, and Preservation of Health, the most temperate Baths.[6]

Aristotelian qualities – hot, cold, moist and dry – must be balanced in the patient by the application of external remedies (such as bathing) or internal remedies (such as draughts or diet), which were also understood in terms of qualities. This is the ancient part of his practice. His modernity lies in his insistence on an epistemology founded on sense perception. For example, Floyer believed that the 'Virtue' of a remedy can be discovered in part by its taste:

> The Astringent Taste discovers a Cooling, Condensating, Repelling, Obstructing Quality.
>
> The Acid attenuates, incides without Heat, and opens and repels.
>
> The Acrid heats, attracts, discusses.
>
> Bitters cleanse, open, attenuate thick Humours, without manifest Heat.
>
> Watery Tastes cool; thicken, obstruct, mortifie, or stupifie.
>
> Salt Tastes dry without great Coldness or Heat, astringe; they preserve from Putrefaction.
>
> Sweet Tastes concoct; mollifie, rarifie, or ferment.
>
> Oily Tastes moisten; mollifie and ease Pain.[7]

The sophistication of Floyer's categorisation of tastes indicates his expert training. Bitter, acrid, acid, astringent, they must all be distinguished in character and degree. His trained senses are the bedrock of his empiricism.

Dr Thomas Guidott of Bath replied to Floyer with the pamphlet *An Apology for the Bath.*[8] Comparing the two authors, we see limits to Floyer's modernism, for he is quite sceptical and distrusting of instrumentation: 'I should never trust the *Microscope* [. . .] All the Advantage by *Microscopes* is, to confirm our other Senses, and help them to discover the Figures of the Mineral Salts which affect the Taste.'[9] Whereas Guidott writes in defence of instrumentation:

> If we had not had better helps than the bare use of our senses, we had not had that distinct knowledge of the mineral waters in our own land we now enjoy, and we owe in good measure to [. . .] the late [. . .] Mr. Robert Boyle.[10]

Microscopes, air pumps, thermometers, clocks – the paraphernalia of the laboratory – were, by the late seventeenth century, acquiring an increasingly secure purchase in medical enquiry, even if, for Floyer and like-minded doctors, they weren't actually much use (yet). The progress of instrumentation in understanding disease and the *materia medica* continued and continues unarrested in our own time. According to a recent commentator, in the early twentieth century we reached a point where 'physicians' senses gradually came to be *replaced* by machine based technologies'.[11] Not augmented, 'replaced'. Calibration, data, mechanism and measurement supplant direct, trained, expert perception. The idea is that they are more reliable and ultimately more 'useful'.

Tourist 1

In Floyer, we have an example of expert practice reliant on a trained sense perception and supported by a tradition that honours such perception and places it in a matrix of understanding. What then of lay perceptions?

Celia Fiennes (1662–1741) was a woman from a wealthy family who travelled extensively in Britain during the last two decades of the seventeenth century. We have a record of her tours in the form of her diaries.[12] Among her diverse observations, she gives considerable space to mineral waters, for example at Harrogate:

> the Stinking Spaw, not improperly term'd for the Smell being so very strong and offensive that I could not force my horse near the well [. . .] The water is furr'd with a white Scum over it, notwithstanding it rises from the Spring very clear [. . .] the taste and Smell is much like Sulphur, tho it has an additional offensiveness like Carrion or Jakes [. . .] its a quick purger and good for all Scurbutick humours – I drank a quart in the morning for two days.[13]

As Fiennes writes, she analyses and categorises (like our doctor) by smell, sight, taste and effect ('a good purger') and builds up a considerable body of lay expertise as she goes. Further observations show her relating the properties of a water to the mineralogy of its location. Thus at Canterbury:

> the taste of the spring in this town seemes to be from a mixt soyle and bears a likeness to the Sulphur Spaw at Epsome and the iron Springs too which are at Tunbridge: what its operation is I cannot tell tasteing halfe a glass of it which I did not like.[14]

She also reports 'experiments' to determine the properties of the water: at Durham, silver could be turned copper colour through immersion in the water;[15] at Tunbridge, she reports of 'weighing the water by the Well and carrying then to the middle of the Walks' to weigh it again and find it lighter. This last experiment was devised and executed in support of the idea that the water loses it 'spiriteous' quality as it is transported away from the well. 'Spirit' was a quality much prized by Fiennes – 'I like no water that rises not quick and runs off apace; that must have most spirit and good off the mincral it comes from'.[16] It would appear that the experiment was a fine example of spurious instrumental 'measurement' being used to support a traditional perception of a quality invisible to the instrumental trial adopted. Direct perception of qualities was beginning to lose its epistemological credibility, even outside the specialised cultures of experimental natural philosophy and medicine. The measurers here start to get the upper hand.

In seventeenth-century Britain, there was for the most part no piped water supply. Where such a novel innovation did exist (parts of London), the supply was intermittent at best.[17] There was certainly no treatment of water before the point of domestic delivery; there was no urban sewerage, other than a certain amount of surface-water drainage. People relied for their water supplies on wells, springs, rivers and cisterns. Even a well-constructed system of urban wells is sufficiently novel to prompt a comment from Celia Fiennes.[18] How then did people decide whether or not the water would be safe to drink, particularly when travelling? The issue must occasionally have been acute. Local knowledge must have been important, but so too must an everyday expertise, developed with the support of a culture in which individual sense perception of the quality of waters was still honoured and routinely practised, in both lay and professional cultures.

Poison

I turn now to the twentieth century and the tragic incident of poisoning at Camelford, Cornwall, in 1988. Owing to mismanagement of a water treatment site, twenty tons of aluminium sulphate were accidentally dumped into the wrong tank at the town's reservoir, poisoning the water supply of the entire district with intolerable quantities of toxic metals. According to press accounts, the water was quite obviously unusual and repellent to both touch and taste. However, over a period of two days, the water authority gave out the advice to concerned callers that they should continue to use the water as normal – mix it with orange juice. Many followed this advice and consequently became ill, in some cases alarmingly and chronically ill.[19]

The press coverage of this incident is divided between, on the one hand, issues of blame, compensation and the failure of a system and, on the other hand, human interest stories in which the 'victims' of the disaster are portrayed (rightly) as betrayed by a system designed to protect them. There are questions over management of the site where the pollutant was dumped; quality control and monitoring of the water; training of the switchboard operators; and ultimate culpability. A question that is *not* asked is why, given the evident repulsiveness of the water, so many people persisted in drinking it? Why did so many people trust the advice of the authorities rather than their own perception and judgement? The silence on this topic is surely an indicator of one overriding, if rather anxiously held, truism of our time: where quality control is in place, there should be no need to rely on our own perception and judgement. Unmediated perception of quality is unreliable. Only scientific analysis and systematic technological control can deliver safety and a basic, acceptable standard of product.

I do not wish to denigrate the importance of properly managed and treated water. We could not now live without it. Neither do I wish to exaggerate the limits of a direct, unmediated perceptual practice such as that of Celia Fiennes. What I *do* wish to point out is the way in which technological systematisation stands between us and our perception of the world, and in doing so leads to a situation where little or no support is given to the development and practice of perceptual skills that the control system supplants. If quality control delivers mediocrity, it also delivers the perfect environment for progressive cultural anaesthesia. To develop particular perceptive capacities, we need a shared culture in which the object and purpose of those perceptions are understood and practised.

Tourist 2

In researching this paper, I decided to pay a visit to Grimshaw's Bath Spa, and to do so in the role of *tourist*. No stalking around the building fully clad with sketchbook and camera. I wanted to bathe. As part of the experience, I booked a 'Mineral Water Bath', which turned out to be more massage than bath. The session gave me an opportunity to engage in conversation with my masseur, 'Tom'. Tom viewed himself as a serious therapist with much to offer. He expressed the hope that eventually the Thermae Bath Spa might become available for use by hospital patients from the nearby NHS Royal Mineral Water Hospital. He observed with chagrin that he has few repeat customers as, in order to access his services, a visitor has to pay not only for a treatment but also for a spa session. For Tom, the possibilities of massage and hydrotherapy go beyond recreational pampering. But the Thermae Bath Spa, while acknowledging the history of medicinal use of the waters ('The thermal waters were used for many years in Bath to treat rheumatic and muscular disorders, as well as skin ailments and respiratory problems'),[20] is set up to deliver 'leisure consumerism' rather than 'healing practice'. As its promotional material explains: 'Thermae Bath Spa is a day spa where you can bathe in the warm, natural mineral-rich waters and choose from a range of spa treatments designed to ease the body and soothe the mind'.[21] Some of the treatments listed – shiatsu, acupressure – are frequently used as serious therapeutic medical practices. Here, however, they are presented as merely 'soothing' and 'easing'. The overall tenor of the spa is leisure and tourism, not healthcare. 'Caviar Face Peel' and 'Yummy Tummy' set the tone.[22] There is a radical discontinuity between what Tom has to offer in the way of 'treatment' and what the Royal United Hospital NHS Trust dispenses. Tom doesn't have a laboratory.

But what of the bathing experience itself? In the New Royal Baths, the water, which comes from the Hetling spring, is the colour of a Badoit mineral water bottle. It is cooled from its natural temperature of 45°C to 33°C, the perfect temperature for floating, which is too hot for swimming. There is no trace of iron colour or the salty accretions that characterise the untreated water of the neighbouring King's Bath and Roman Baths, which are now a part of the neighbouring Roman Baths Museum. It all feels clean. There is no sense of earthiness, just a whiff of chlorine. The quality of the water is quite different from ordinary tap water. It has a buoyancy that seems to go beyond mere specific density. Is this what Celia Fiennes calls 'Spiriteousness'? I floated. The water is pushed into currents by strong jets, which propel you along anti-clockwise in the lower bath (Minerva) or straight across the pool in the roof-top bath. By the time I emerged from the steam rooms (which clang and echo like a ship), I was so immersed in wateriness and steam as to be almost overloaded with sensation. It's a good day out.

Architect

'We are not a theoretically driven practice – we like to build', says Nicholas Grimshaw.[23] However, scrutiny of the publications of Nicholas Grimshaw and

Partners uncovers a set of persistent theoretical concerns that inform the work of the practice: expressivity of structure; response to place; transparency; and, in Grimshaw's own words, 'a fundamental belief in people's right to manipulate their own surroundings'.[24] This last aspiration is often manifest in the works of the practice as a highly visible concern with flexibility and demountability, which is carried through into the detailed design of kit-like building systems. Good examples of this are the Waterloo International Terminal in London (1993) and the Eden Project in Cornwall (1996–2001). An early example is the Herman Miller Factory in Bath – an essay in panel demountability. A series of photographs of this last project is published in *Architecture, Industry and Innovation: The Early Works of Nicholas Grimshaw and Partners*. It celebrates the systematic flexibility embodied in frame and panel: 'There was no compromise. Solid panels can be changed for glazed units, doors can be moved, even the position of the recessed courtyards can be changed.' The captions to the photographs tell a story: 'Nick Grimshaw checking a prototype grp panel for flatness'; 'Sequence showing maintenance personnel removing a grp panel'; 'three out of four of the building's elevations face public spaces. *Here local schoolchildren happily race home past the building*'.[25] The framing and presentation of this suite of photographs proclaim a message: we will design and monitor a well-tuned system that gives the occupants freedom to 'manipulate their own surroundings'.[26] From this will arise involvement, participation and spontaneous enjoyment and action: freedom!

12.3 *(opposite)*
New Royal Baths entrance front

The site of the New Royal Baths complex is highly protected by various levels of legislation and interest: the entire city of Bath is a World Heritage Site; the Cross Baths are a World Sacred Site; the Grimshaw intervention is hemmed in on all sides by listed buildings. In this context, high-specification 'local' finishes – Bath stone and glass – and external formal reticence are almost a given; planning control would require it. Grimshaw has nonetheless managed to build a modern structure with no hint of pastiche. The building either hides behind pre-existing eighteenth-century structures at the front or reflects them in a glass wall at the side and back. The language is one of technical solution, kit, system – a rhetoric of flexibility (Figure 12.3). Given the practical constraints of delivering high volumes of carefully filtered and cooled natural spring water to the bathing facilities, and given the enormously high costs of the project and the physical and legislative constraints of the location, the room for actual demountable flexibility is slight in the Baths complex and, anyway, why would it be required? So it is slightly strange to find the rhetoric of the demountable emerging so strongly in moments such as the entrance canopy to the roof-top pool (Figure 12.4); the sci-fi purity of the steam-room capsules (Figure 12.5); and, most strangely, in the detailing of the Bath stone wall that encases the cubic volume of accommodation hovering above the principal pool (Figure 12.3). The fairly traditional ashlar stonework is jointed to look like a demountable panel system.[27] The building, in a quiet way, declares what many of Grimshaw's buildings shout: get a good flexible system in place and you will be free to enjoy.

As I bobbed around in the roof-top pool, an adjacent female bather, daring to notice her own responses to the water, observed with a mix of hesitancy and defiance that she felt the water to be good for her sinuses. The boys sharing the eucalyptus steam room pronounced the aroma to be 'subtle'. All of which is

12.4
One of the four steam pods, before occupation, before steam

12.5
The roof-top pool

a start. It's not that, even with our relatively deskilled senses, we can't see, feel, smell the water. It's just that we are not quite sure what it's all for. Good for our health? Perhaps? Given that we can't really agree what, exactly, health is (a functional capacity, a state of psychosexual balance, a plenitude of being?), this is a rather difficult question to answer.[28] Grimshaw's architecture – with its deferential framing of occupant activity set within a structure representing systematic flexibility – is perhaps the perfect setting for such ambivalent ponderings. As Nicholas Grimshaw says, 'when I sit on the rooftop bath looking out over the City I think I will feel that I have done something for the City'.[29] The architect aspires to be elevated above the town to observe it in removed perspective. It would be possible to interpret this also as a part of the discourse of removal from the immediacy of bodily experience. However, here, the viewer is the bather, immersed up to the chin in spirituous thermal waters. This particular view is certainly not a disembodied experience.

Notes

1 Donald Insall Associates acted as experts in historic building conservation on the project.

2 www.thermaebathspa.com/index.html (accessed 27 May 2009).

3 Roger Rolls, *The Hospital of the Nation: The Story of Spa Medicine and the Mineral Water Hospital at Bath*, Bath: Bird, 1988. Henry William Freeman, *The Thermal Baths of Bath: Their History, Literature, Medical and Surgical Uses and Effects; Together With the Aix Massage and Natural Vapour Treatment* (S.l.) London: Hamilton, Adams, 1888.

4 www.thermaebathspa.com/index.html (accessed 27 May 2009).

5 Sir John Floyer, *An Enquiry Into the Right Use and Abuses of the Hot, Cold, and Temperate Baths in England, etc.*, London: R. Clavel, 1697.

6 Floyer, pp. 10–11.

7 Floyer, pp. 5–6. Much of Floyer's literary output was devoted to classifying medicines for their usefulness in different conditions, according to their tastes and smells. D.D. Gibbs, 'Floyer, Sir John (1649–1734)', *Oxford Dictionary of National Biography*, Oxford: Oxford University Press, 2004. Available online at www.oxforddnb.com/view/article/9775 (accessed 20 May 2009).

8 Thomas Guidott, *An Apology for the Bath, Being an Answer to a Late Enquiry (by Sir J. Floyer), into the Right Use and Abuses of the Baths in England*, I, London, 1705.

9 Floyer, p. 38.

10 Guidott, 'An Apology, Postscript', quoted in Henry William Freeman, *The Thermal Baths of Bath: Their History, Literature, Medical and Surgical Uses and Effects; Together With the Aix Massage and Natural Vapour Treatment* (S.l.), London: Hamilton, Adams, 1888, pp. 119–20.

11 Merriley Borell, 'Training the Senses, Training the Mind', in W.F. Bynum and Roy Porter (eds), *Medicine and the Five Senses*, Cambridge: Cambridge University Press, 1993, pp. 244–61 (p. 245).

12 The most comprehensive and scholarly edition of the diaries is: Celia Fiennes, *The Journeys of Celia Fiennes*, Christopher Morris (ed.), London: The Cresset Press, 1947. Celia Fiennes, *The Illustrated Journeys of Celia Fiennes 1685–c1712*, Christopher Morris (ed.), London: Macdonald, 1982, contains additional commentary on context.

13 Fiennes, 1947, p. 80.

14 Fiennes, 1947, p.125. She records similar observations on mineral waters in at least fourteen other locations, including both famous spas – Bath, Tunbridge, Epsom, Harrogate, Scarborough – and those less well known – Barnet, Rotherham, 'Astrop'.

15 Fiennes, 1947, p. 216.

16 Fiennes, 1947, p. 125.

17 Nicholas Barton, *The Lost Rivers of London*, London: Historical Publications, 2000, especially 'Water Supply', pp. 81–6.

18 Fiennes, *The Illustrated Journeys*, p. 97.

19 The scope of my study does not include any detailed understanding of the Camelford incident, and I apologise for any offence or inaccuracy in my presentation of the events. For typical news

coverage see: http://news.bbc.co.uk/1/hi/health/455871.stm, Thursday, 23 September 1999, published at 18:09 GMT 19:09 UK; and www.dailymail.co.uk/news/article-502442/A-lethal-cover-Britains-worst-water-poisoning-scandal.html (accessed 27 May 2009).

20 During the twentieth century, the waters were used for rehabilitation of soldiers after World Wars I and II, and for treatment of patients with mobility problems and rheumatic disorders. The baths associated with Hetling Spring were closed in the 1970s owing to an outbreak of meningitis that was traced back to the water. See Rolls for a discussion of the medical use of the water; see Freeman for plans of the twentieth-century spa facilities.

21 www.thermaebathspa.com (accessed 27 May 2009).

22 www.thermaebathspa.com/treatments/ourtreatments/atozortreatments/ (accessed 27 May 2009).

23 Nicholas Grimshaw, in the 'Foreword' to Hugh Pearman, *Equilibrium: The Work of Nicholas Grimshaw and Partners*, London: Phaidon, 2000, p. 7.

24 Nicholas Grimshaw, 'Autobiographical Note', in Colin Amery, *Architecture, Industry and Innovation: The Early Works of Nicholas Grimshaw and Partners*, London: Phaidon, 1995, p. 16. For Grimshaw in general, see also: Rowan Moore (ed.), *Structure, Space and Skin: The Work of Nicholas Grimshaw & Partners*, London: Phaidon, 1993; Nick Grimshaw, 'Buildings as Living Organisms', video recording, London: PAV 21–9002, Pidgeon Audio Visual for Sussex Video, 2–6 Foscote Mews, London W9 2HH.

25 Amery, *Architecture, Industry and Innovation*, pp. 74–87.

26 See note 22.

27 For detail of this wall, see 'In Detail', *Building Design*, 4 August 2006, p. 17.

28 For a definition of 'health' from outside the modern medical culture see Zkixing Wang, 'A Higher Standard of Health', CD audio recording, London: Chinese Heritage Institute, May 2005.

29 Nicholas Grimshaw, 'Bath Spa, 1997–2001', in Pearman, pp. 68–77 (p. 68).

Chapter 13

The qualia of architecture

On Sullivan, Szarkowski, autobiographies and ideas

Hugh Campbell

John Szarkowski's book *The Idea of Louis Sullivan*, first published in 1956, consists of ninety of his black and white photographs of Sullivan's buildings. Among the most striking is an image of the Guaranty Building in Buffalo, New York, seen head on, from an elevated viewpoint, the façade almost filling the frame. Although all twelve bays and ten storeys of the main body of the building are included in the photograph, the assymetrical composition suggests that the rhythm of window and frame extends outwards and downwards, and that we are simply seeing part of some larger order. The flat orthogonality of the image accentuates the order's highly abstract character, which is then undercut by the glimpses of life seen through the windows. Lamps, blinds and occasional figures punctuate the rhythm of the structure. Sullivan's architecture is presented as a framework within which the individual dramas of life can be played out. We are invited to reflect on the nature of the relationship between ourselves and the built environment that surrounds us – on, for instance, the two men standing at an open window – and on the specific qualities lent to this moment by the architecture that frames them (Figures 13.1 and 13.2).

From its earliest moments to its dying seconds, each of our lives constitutes a continuous and ongoing encounter with the world. It is by means of our ceaseless meetings with, and reactions to, other objects and organisms that we establish, define and explain ourselves. And it is by the same means that we come to name and know those objects and organisms that constitute our familiar world. There is, in other words, a reciprocal relationship between consciousness and situation, between who we are and where we are. In *The Cultural Psychology of Self*, Ciaran Benson suggests that a process of spatial orientation and differentiation is in fact central to the very creation of the self.[1] We begin to know who we are through our understanding of where we are. But, as Benson acknowledges, it is difficult to infer a simple causality from what is essentially a circular process. The self may not initially be spatial, but once we use spatial terms to understand it, those become the terms by which we chart its development. Metaphor enters in from the outset, and governs all subsequent understanding. As Georg Lakoff and Mark Johnson explain:

> In all aspects of life . . . we define our reality in terms of metaphors and then proceed to act on the basis of the metaphors. We draw inferences, set goals, make commitments and execute plans all on the basis of how we in part structure our experience, consciously and unconsciously, by means of metaphor.[2]

In particular, the metaphors of inside and outside and all their correlates become so deeply embedded as to be both inescapable and endlessly useful to our understanding of conscious experience – both our own and that of others.[3] And the moment we begin to interact with our surroundings, further complexities become overlaid upon our sense of the self's location. It is at this meeting point between internal and external worlds that the English psychologist Nicholas Humphrey has positioned his most recent exploration of consciousness. In his brief, eloquent book *Seeing Red*, Humphrey's main focus is on the problem of 'qualia'.[4] Qualia may be described as the properties of sensory experiences. In efforts to isolate the particularities of conscious experience, researchers of consciousness often focus on how particular sensations – the feeling of pain, the perception of the colour red – occur in the brain.[5]

From the outset, Humphrey draws a fundamental distinction between perception and sensation. The two are almost independent, and can sometimes operate without each other. But usually, sensation can be seen as the registration of perception – the means by which we become aware of having perceived (of the physical process of light striking the optical nerve, for instance). For Humphrey, sensation is essentially the evolved version of a feedback loop. And it is what creates the self, by generating an awareness of experience alongside the immediacy of experience itself.

Humphrey expands the argument in evolutionary terms, explaining that, as the processing of perception begins to happen at a remove from the site of its generation (that is, as central brains begin to evolve in organisms), there is a concomitant development away from immediate response towards reflexive perception. A gap opens between action and reaction. It is in this gap that sensation and, hence, self-awareness emerge. Where once there was only the bare capacity

13.1

Louis Sullivan, The Guaranty (now Prudential) Building, Buffalo, New York, 1894–5. Photograph by John Szarkowski supplied courtesy of the artist and the Pace/MacGill Gallery, New York

13.2

Louis Sullivan, The Chicago Stock Exchange (later 30 North LaSalle Building), 1893–4, demolished 1970–1. Photograph by John Szarkowski supplied courtesy of the artist and the Pace/MacGill Gallery, New York

to react to external stimuli, there is now an added ability to be aware of and control that action. But even as emerging consciousness – in the form of sensory activity – internalises and privatises reactions, its essential characteristic as something active and responsive remains. Humphrey sees sensation as always active rather than passive: he 'puts sensation within the sphere of agency, on the production side of the mind rather than the reception side'.[6]

Central to Humphrey's explanation is the idea that consciousness evolves from the outside inwards, rather than, as might more intuitively be thought, from the inside outwards.[7] As reactions to external stimuli evolve into internal states of reflection, consciousness perpetuates the feeling of there being a slight distance between the organism and what happens to it. Even when we are examining our own actions and mental activity, we replicate the manner in which we examine the world beyond.[8] For Humphrey, so-called 'mirror neurons', which 'link the *observation of someone else having a sensation to the execution of a similar sensation oneself*', are evidence of the outward-directed nature of consciousness.[9] If the human brain has produced consciousness as a kind of hyper-evolved version of our reactions to other organisms and environments, it is hardly surprising that it should retain a pronounced capacity for what Humphrey chooses to define as 'empathy'. He quotes Nietzsche on the subject:

> To understand another person, that is to imitate his feelings in ourselves, we [. . .] produce the feeling in ourselves by imitating with our own body the expression of his eyes, his voice, his walk, his bearing. Then a similar feeling arises in us in consequence of an ancient association between movement and sensation. We have brought our skill in understanding the feelings of others to a high state of perfection and in the presence of another person we are always almost involuntarily practicing this skill.[10]

Discussions of empathy – in German, *Einfühlung* – were prevalent in German intellectual culture in the late nineteenth century. However, the term was primarily used, not to examine consciousness itself, or the relations between people, but rather to understand the aesthetic experience of works of art and architecture, a school of thought usually referred to as 'empathy theory'. As Mitchell Schwarzer explains: 'In empathy theory, the source of pleasure resides neither in the object nor in the subject, but in the relationship of the consolidating perception between object and subject'.[11] Writers such as Theodor Lipps, August Schmarzow and Heinrich Wölfflin sought to concentrate their discussions of works of art on the detailed analysis of how they were perceived by the human subject. Partly, this was in an effort to counter explanations of aesthetics based solely on their inherent content and narrative, and partly it was in response to the emerging field of psychology, pioneered by figures such as Johann Friedrich Herbart. Building on Kant's separation of perception and cognition, Herbart 'argued that the human mind is a simple and unchanging entity at birth which, when stimulated by the world, responds with ideas [that] compose the statics and dynamics of mental life'.[12]

In empathy theory, the ideas constantly forming and reforming in this active mind are explained as attempts to conceptualise the perceived world in the human image. As Heinrich Wölfflin saw it:

> Forms become meaningful to us only because we recognize in them the expression of a sentient soul. Instinctively we animate each object [. . .] We read our own image into all phenomena. We expect everything to possess what we know to be the conditions of our own well-being. Not that we expect to find the appearance of a human being in the forms of inorganic nature: we interpret the physical world through the categories that we share with it. We also define the expressive capability of these other forms accordingly. *They can communicate to us only what we ourselves use their qualities to express.*[13]

Returning now to Szarkowski's *The Idea of Louis Sullivan*, it is evident from its content and even from its title that the book shares with the empathy theorists a conviction that architectural form and mental life have common patterns. This conviction is also central to the book that Szarkowski's title is clearly intended to evoke – Sullivan's curious and compelling account of his own life, *The Autobiography of an Idea*.

The Autobiography of an Idea dwells extensively on Sullivan's childhood and early training and says relatively little about his mature work as an architect. The book is narrated entirely in the third person ('Louis'). This has the disconcerting effect of distancing and 'objectifying' all the reported events and emotions, but is completely in keeping with the pattern of the book, in which Sullivan's character – his autonomous self – is formed through a series of encounters, first with physical and natural things, and later with abstract orders and systems. In each case, a new stage of self-awareness is marked through Sullivan's understanding of the discrete nature of the things to which he forms strong attachments. Early on, this happens with the plants he is fond of tending:

> How often on hands and knees had he peered and gazed long [. . .] indeed worshipped them in friendship until he seemed to feel them grow: that they were of this world and yet not of his world, that they seemed to live their own lives apart from his own life.[14]

And then later, with an elm tree:

> But the lovely elm was his infatuation – he had adopted her at first sight, and still gazed at her with a sweetness of soul he had never known. He became inflitrated, suffused, inspired with the fateful sense of beauty. He melted for an instant into a nameless dream, where he saw she was sufficient unto herself, that like his garden plants she lived a life of her own, apart from his life. Yet they both lived in the same big world – they both, for the moment, stood in the same green field. Was there nothing in common? Did she not know he was there?[15]

Sullivan subsequently goes through the same process of absorption and separation with the city of Boston, with his beaux-arts training, with Chicago's burgeoning capitalist society. In each case, coming to terms with external phenomena clarifies Sullivan's sense of his own self, and the experience galvanises him towards more concerted and deliberate action. The pattern recalls Humphrey's description of a consciousness created from the outside in, and the notion – as elaborated from Kant by Johann Herbart – of the conscious self developing through the shaping of experience into ideas. It is not clear whether Sullivan knew the work of Herbart directly, but he certainly absorbed a lot of nineteenth-century German philosophy, the influence of Nietszche clearly evident in passages such as the following:

> Louis saw power everywhere; and as he grew on through his boyhood, and through the passage to manhood, and to manhood itself, he began to see the powers of nature and the powers of man coalesce in his vision into an IDEA *of power*. Then and only then he became aware that this idea was a *new idea* – a complete reversal and inversion of the commonly accepted intellectual and theological concept of the Nature of man. That IDEA which had its mystical beginning in so small a thing as a child's heart, grew and nurtured itself upon that child's varied consistently continuing and metamorphosing experiences in time and place.[16]

Much has already been written on the influence of American thinkers and writers on Sullivan. He drew on Thoreau, on Emerson, on Willam James and, above all, on Whitman, to concoct his own blend of American 'trascendental pragmatism', summed up in the book's final chapter:

> And thus the natural man ever enlarges his range of beneficence. His life experiences are real. He reverses the dictum 'I think: Therefore I am.' It becomes in him, *I am: Therefore I inquire and do!*
>
> It is this affirmative 'I AM' that is man's reality.[17]

This is the idea whose autobiography Sullivan writes. The intense, insistent manner in which he intertwines his personal development with his thoughts on architecture and society is faithfully reflected in Szarkowski's book, in which photographs are interleaved with passages from Sullivan's writings and those of his contemporaries, admirers and sources of inspiration. Included are extracts from Walt Whitman's *Leaves of Grass*, from Frederick Turner's *The Significance of the Frontier in American History*, from Thorsten Veblen and from Frank Lloyd Wright's own autobiography. No direct connections are suggested between words and images – rather they are two parallel streams that, between them, encompass the creation and the subsequent life of these buildings. Szarkowski's camera acts to mediate this relationship.

Szarkowski's project began in 1951, when he found himself teaching in Buffalo and began photographing the Guaranty Building:

> As I began to work I found, to my own surprise, that I was seeing this building not with the decorous disinterest with which a photographer is supposed to approach a work of formal architecture, but as a real building, which people had worked in and maimed and ignored and perhaps loved, and which I felt was deeply important. I found myself concerned not only with the building's art-facts but with its life-facts. (Louis Sullivan had claimed they were the same.) This concern began to show in the photographs, and the idea grew.[18]

Over the following four years, Szarkowski made a large series of deliberate, painstaking images, mostly with a medium-format Graphic View camera.[19] The architecture and the photographs share a calm, powerful clarity. The dark tones and sharp shadows of the photographs reinforce that feeling of weight and measure particular to Sullivan's architecture. The deep, crisp focus serves to animate the heavy vegetal ornamentation (Figure 13.3). Framing and composition are used to produce a rich interplay between architectural intention and lived experience. Szarkowski balances the visual relationships in the gridded glass of his viewfinder, conferring coherence and order on the seemingly disparate elements of the scene. Thus, the photographer's eye (to use the title of one of Szarkowski's later books) becomes analagous to, and offers an extension of, Sullivan's own methods of intellectual development:

> To a photographer the world consists of an infinite number of vantage points – places to stand – of which very few are altogether satisfactory. The photographer's goal is simple in principle and seemingly modest in ambition: it is to find the place and moment from which some interesting aspect of the world can be converted into a photograph that will be both clear and lively.
>
> Alas, when we remove from the world its space, texture, smell, temperature, sound, and often color, the aspect that remains is seldom either clear or lively. We aim for clarity and generally get stasis; we hope for liveliness and too often get what is merely messy.
>
> Nevertheless, occasionally photographs are made that are good to look at and unfamiliar (clear and lively), and from those pictures we may learn something.
>
> [. . .] Such occasional small blessings accumulate in a photographer's mind, not in a form conscious enough to be called ideas, except in the most inchoate form, but as a growing awareness of the pictorial possibilities of the world – of some of the ways which the visual data of the world form clusters – that can be recognized in very different circumstances.[20]

As if in anticipation of this affinity with the camera, in *The Autobiography of an Idea* Sullivan draws attention to the fact that his name in Irish – O'Suilleabhain – means One Eye, and he imagines an ancestor, the Prince with One Eye, who

13.3
Louis Sullivan, The Schlesinger-Meyer (now Carson Pirie Scott) Department Store, Chicago, 1899–1904. Photograph by John Szarkowski supplied courtesy of the artist and the Pace/MacGill Gallery, New York

13.4
Louis Sullivan, The Chicago Auditorium, 1886–9. Photograph by John Szarkowski supplied courtesy of the artist and the Pace/MacGill Gallery, New York

'must have seen much'.[21] And in the later edition of his book, Szarkowski overlays further autobiographical resonances, recounting how he came to show these photographs to Frank Lloyd Wright – Sullivan's apprentice and self-appointed heir – and then subsequently to Walker Evans, Szarkowski's own direct photographic forebear and future friend.[22] *The Idea of Louis Sullivan* folds closely together the autobiographical experiences of both men – Szarkowski's own adventure in finding his photographic voice and Sullivan's architectural development through words and buildings. Central to both is the notion of the 'idea', which for Sullivan and Szarkowski – as it had for Herbart – constitutes a kind of willed concordance between the self and the world. But this concord is not reserved purely for the artistic creator. What Szarkowski's luminous photographs also propose is that the encounter with architecture can lend conscious shape to our everyday existence (Figure 13.4).

Qualia, as Nicholas Humphrey proposed, emerge in the process of translation from sensation to perception, producing what he terms the 'thick moment' of consciousness. It is from the steady, unending accumulation of these thick moments that our autobiography is written. The neurologist Antonio Damasio refers to consciousness as the continuous enactment of 'the autobiographical self'.[23] But autobiography is something more than the mere accumulation of experience – it is experience deliberately shaped, told and retold as a story to ourselves and to others. As Sullivan's own autobiography suggests, we become ourselves only insofar as we develop the capacity to differentiate, define and describe that self. From this, it might be concluded that qualia themselves can only be said to exist to the extent that they can be described and communicated. In *Consciousness Explained*, Daniel Dennett takes the view that qualia are like versions of Richard Dawkin's 'memes' – ideas that can become transmitted from one human to another so that they eventually acquire a quasi-genetic status.[24]

The photographs in *The Idea of Louis Sullivan* are certainly intended to transmit something of the architect's creative process.[25] More importantly, though, they allow us to trace in reverse the path from image to lived experience, communicating completely and precisely the encounter with Sullivan's buildings – their art-facts and their life-facts, their role in countless unwritten autobiographies.

Notes

1 Ciaran Benson, *The Cultural Psychology of Self: Place, Morality and Art in Human Worlds*, London: Routledge, 2001.

2 Georg Lakoff and Mark Johnson, *Metaphors We Live By*, Chicago, IL: University of Chicago Press, 2003, p. 158 (first published 1980).

3 Here, for instance, is Helen Mirren explaining how she prepared to play Queen Elizabeth II:

> As she was speaking about the role, she suddenly stopped and held her hands up in front of her face. 'I went like this before I did a take,' she said, pulling her palms slowly toward the back of her head, then suddenly swiveling her wrists and thrusting her hands forward, like beams of light, beside her eyes. 'She's way back inside herself,' she explained. 'Her personality, her intelligence, everything is way back. Then she's steadily looking out, as if through a porthole, with this incredibly nonjudgmental, confident gaze.'

John Lahr, *Command Performance: The Reign of Helen Mirren*, *New Yorker*, 2 October 2006. Available at www.newyorker.com/fact/content/articles/061002fa_fact1 (accessed 16 November 2009).

4 Nicholas Humphrey, *Seeing Red: A Study in Consciousness*, Cambridge, MA: The Belknapp Press of Harvard University Press, 2006.

5 Unsurprisingly, there is a wide divergence of views on the nature of qualia, with some, including Daniel Dennett, questioning whether they can be said to have any independent existence at all. See, for instance, the lengthy entry on qualia at http://en.wikipedia.org/wiki/Qualia (accessed 16 November 2009) and the chapter 'Qualia Disqualified' in Daniel Dennett, *Consciousness Explained*, London: Penguin, 1993, pp. 369–411.

6 Dennett, p. 127.

7 In evolutionary terms, the capacity of humans to receive a continuously updated report on our own condition confers clear advantages. Consciousness may be an illusion – a by-product of the perceptual process – but for Humphrey it is a 'deliberate trick' rather than 'an honest error', a trick that, because it allows us to survive and flourish, becomes increasingly part of our genetic make-up.

8 This is particularly noticeable in young children of four and five, who, although speaking in the first person, often report on their own feelings and desires in a surprisingly objective, distanced manner. Some would argue that it is precisely the use of language that produces this 'distancing' effect – that remakes the child as the 'first person singular'.

9 Humphrey, p. 106 (original italics).

10 Frederich Nietzsche, 'Daybreak', in R.J. Hollingdale (ed. and trans.), *The Nietzsche Reader*, 1881, Harmondsworth: Penguin, 1977, p. 156, quoted in Humphrey, p. 104.

11 Mitchell Schwarzer, 'The Emergence of Architectural Space: August Schmarzow's Theory of Raumgestaltung (Spatial Forming)', *Assemblage*, August 1991, no. 15, pp. 48–61.

12 'Introduction', in Harry Francis Mallgrave and Eleftherios Ikonomou (eds), *Empathy, Form and Space: Problems in German Aesthetics, 1873–1883*, Santa Monica, CA: Getty Center Publications, 1994, p. 11.

13 Heinrich Wölfflin, 'Prolegomena to a Psychology of Architecture', in Mallgrave and Ikonomou (eds), *Empathy, Form and Space*, pp. 149–92, p. 152 (original italics).

14 Louis Sullivan, *The Autobiography of an Idea*, New York: Dover Press, 1956, p. 62.

15 Ibid., p. 64.

16 Ibid., pp. 248–9 (all capitals and italics original). The inventory of Sullivan's library contained in David S. Andrew, *Louis Sullivan and the Polemics of Modern Architecture: The Present Against the Past*, Chicago, IL: University of Illinois Press, 1986, pp. 161–70, mentions only one title by Nietzsche – *Also Sprach Zarathustra*.

17 Sullivan, *Autobiography*, p. 265. On Sullivan's intellectual sources, see, for instance, Andrew, pp. 23–74.

18 John Szarkowski, Foreword to 1956 edition, in *The Idea of Louis Sullivan*, London: Thames & Hudson, 2000, p. xvii.

19 A few were taken with a more lightweight Rolleiflex.

20 John Szarkowski, *Atget*, New York: Calloway, 2000, p. 50.

21 Sullivan, *Autobiography*, p. 37. Szarkowski's most enduringly influential book, which has recently been reissued, is *The Photographer's Eye*, London: Thames & Hudson, 2007 (first published 1966).

22 Szarkowski died on 7 July 2007, the day after this paper was delivered at the Quality conference in Cardiff. In his role as curator of photography at the Museum of Modern Art between 1962 and 1991, he was a hugely influential force in American photography.

23 Antonio Damasio, *The Feeling of What Happens: Body, Emotion and the Making of Consciousness*, London: Vintage, 2000, pp. 168–77 and *passim*.

24 Dennett, *Consciousness Explained*, pp. 199–210.

25 As Terence Riley points out in his introduction to the more recent edition: 'Szarkowski's book acknowledges that any real understanding of Sullivan's work on the part of the reader derives from the leap from the image to the idea – in effect, tracing in reverse the path Sullivan himself took.' Terence Riley, Introduction to *The Idea of Louis Sullivan*, 2000, p. xiv.

Chapter 14

The architect's signature as a sign of quality

Paul Emmons

Qualities are signatures, but the signature, the proper name, is not the constituted mark of a subject, but the constituting mark of a domain, an abode.

(Deleuze and Guattari, *A Thousand Plateaus*, 1987)

The whole outward visible World with all its Being is a Signature of the inward spiritual World; whatsoever is Internally, and howsoever in operation is, so likewise it hath its Character externally.

(Jacob Boehme, *Signatura Rerum*, 1651)

A widely accepted sign of quality in the visual arts is the artist's signature, which confirms with a mark an object's authenticity, authorship and completeness. While often overlooked, architects' signatures appear on buildings in every era and yet remain relatively rare. The reason is often assumed to be the indirect relation of the architect to the building object. On the one hand, the builder is more directly responsible for the construction and execution, while, on the other hand, the patron is more in control of the idea and inception.[1] Lucian exemplifies the cunning required for architectural signatures in the story of Sostratos, the architect of the Alexandria lighthouse, which was one of the seven wonders of the ancient world. Sostratos set the reigning king's name in impressive gold letters on the tower, but in plaster, so that over the years, it aged and fell away, revealing underneath the architect's name permanently incised in stone.[2] Even for the signature, the architect autographs a drawing to direct others to inscribe his or her name on the building.

The frequently assumed distinction of architecture from the visual arts because of the lack of immediacy of the architect's hand in making the final work does not hold. Painters and sculptors have long used assistants and shop practices to produce work.[3] Giovanni Bellini's renaissance shop even standardized the master's signature so that others could paint it.[4] The signature – reflecting an intellectual conception rather than a manual execution – reveals that the other arts are much more like architecture, so that the divided agency of the signature can be true for all the visual arts.

The particular way in which a building is signed often betrays the architect's complex relation to the building. By tethering them together through a name, the signature makes the architect integral with, but simultaneously at a distance from, the edifice. This elided identity is expressed by the nature of the signature as manifest in its location, material and manner. In short, the signature expresses particular qualities.

The myriad ways that buildings are signed can best be categorized through Charles Peirce's triad of signs: icon, index and symbol.[5] Iconic signatures are most often architects' portraits integrated with their buildings. Much as renaissance artists inserted themselves within their paintings to gaze out at the viewer, architect Henry Bacon greets visitors to the Lincoln Memorial from the northern mural.[6] The busts of medieval master masons often appear within their great works. Notably, Peter Parler emerges from a column in the triforium of Prague Cathedral.[7] Portrait signatures continue to appear in the twentieth century; for example: Cass Gilbert as a bespectacled caryatid holding the model of the Woolworth Building (New York City, 1913) and, even more playfully, multiple Charles Moore heads spitting water in the Piazza d'Italia (New Orleans, 1978).[8] Perhaps the most direct iconic signature is when the building serves as the architect's burial place. This often happened in church buildings that were designed for burials, and occasionally the architect is interred in a building that was not otherwise intended for the deceased.[9]

The architect's signature, as an inscription of his or her name, is an indexical sign with a direct physical relation to the building. However, it need not be an autograph, requiring only the author's consent and instruction rather than their direct execution of the physical mark. Owing to the coincidence of his name, one of the best-known signatures from the ancient world is: 'L. VITRVVIVS. L. L. CERDO. ARCHITECTVS.'[10] Whether or not he was the design architect, the presence of his name inspired signatures in the Renaissance and among following generations of architects. The architect of St Martin-in-the-Fields (London, 1726), for example, signed 'IACOBO GIBBS ARCHITECTO' on the front pediment entablature.[11]

A symbolic signature is a device or mark other than a name that in some way stands for the architect, just as medieval masons' marks were sometimes used.[12] Perhaps the most famous signature symbol is to be found in Pliny's report of the architects Saurus and Batrachus, who built a temple in Rome and, upon being refused the privilege of inscribing their names, carved the rebus of each name, a lizard and a frog, into the spirals of the Ionic column capitals.[13] Beyond identification, symbolic signatures can act as talismans to invoke heavenly protection, most clearly when signatures are not meant for mortal eyes.[14]

It is never unpremeditated how and where architects sign their work. In a manner that is always significant, it is testimony to the mediated relationship between the creator and the creation. This is especially true in art museums, where the idea of the architect as artist is often paramount. The qualities of three signatures from the twentieth century will be explored here.

The speaking building

ERNEST·FLAGG·INV
MDCCCXCIII

The Corcoran Art Gallery (Washington DC, 1897) is signed by Ernest Flagg in the lower right-hand corner of the main front façade, in much the way an artist signs a finished canvas (Figure 14.1).[15] The dramatically blank wall dominating the front façade allows the display of paintings inside, so that the artists' signatures within are mirrored by Flagg's building signature on the outside of the wall. That Flagg's name appears below the string course of the main façade is directly reminiscent of the early Renaissance tradition of signing paintings on the frame. Giotto, whose name appears on the frieze of the Corcoran as one of eleven 'distinguished artists' and who was *capomaestro* of Florence's *Duomo*, signed his paintings on the

14.1
Architect's signature on front façade, Corcoran Art Gallery, Ernest Flagg

frame.[16] Signed on the frame, the mark does not intrude into the integrity of the artwork, and it provides a link between the worlds of the viewer and the artwork. That the artists' names are inscribed in a classical roman typeface, while Flagg's name is in sans serif letters, heightens this differentiation and propels Flagg towards the viewer as the mediator of ancient tradition.

Flagg's signature was approved by the building committee, as architects rarely sign a building without the patron's permission.[17] Just two years earlier, McKim, Mead and White had secretly arranged the rows of names of famous authors and artists on the façade of the Boston Public Library to insert the firm's name as an acrostic. The offending order of names was as follows:

MOSES	MOZART	WREN
CICERO	EUCLID	HERRICK
KALIDASA	AESCHYLUS	IRVING
ISOCRATES	DANTE	TITIAN
MILTON		ERASMUS

The angry public outcry against this 'self-promotion' led to the firm being compelled to reconfigure the order of the names by incising them anew in order to remove their own names.[18]

After Flagg's name in his signature is 'INV' for the Latin *invenit*, designating the inventor or designer. He was aware that 'INV' is used in printmaking to identify the artist of the original drawing and contrasts with 'FECIT' or 'SCULP', which describes the person who executed the block or plate.[19] Through this division between ideation and realization, Flagg was able to distinguish himself from the mason who cuts the architect's drawn design into stone. The building inscription directly transfers Flagg's signature from a similar location on his ink-wash rendering of the front elevation, closely paralleling drawing and building.

Flagg's attendance at the École des Beaux-Arts is the likely influence for his predilection to sign buildings. The signing of private and commercial buildings was relatively common in Paris.[20] The architects were likely propelled by the emerging nineteenth-century understanding of the architect as fine artist, as evidenced by the reorganization of the École, uniting architecture with the visual arts.[21] Flagg used Garnier's Opera, which is profusely signed, to exemplify his idea of architectural quality as fine art:

> A great work of architecture, like any other great work of art is chiefly valuable for the seal of personality which it receives from its author [. . .] Take the Paris Opera House as a case in point. The work is stamped all over with the personality of Charles Garnier, and although opinions may differ in regard to matters of detail, all will agree that this author is a man of genius; his production will always rank as a work of art.[22]

Garnier's obscured signature in the Opera may have provided a model for, among others, Louis Sullivan, also once a student in Paris, who signed his initials into the ornament of the Carson Pirie Scott building (Chicago, 1899).[23] Flagg concluded, in 1908:

> I am heartily in favor of signing buildings. I think it ought to be done in all cases except where the architect is ashamed of his work. If a building is a work of art, I do not see why it should not be signed by the artist, just as any other work of art is signed. I have done it myself in a number of instances, and I hope it may soon become a universal practice.[24]

The idea of art for art's sake as an expression of artistic genius was particularly influential at this time, as demonstrated by the architects' signatures.[25]

Reification of the name

ARCHITECT
I.M. PEI & PARTNERS

In the East building of the National Gallery of Art (Washington DC, 1978), I.M. Pei's name is signed on a central triangular column in the atrium (Figure 14.2).[26] Pei recalled that the design began with drawing a line through the trapezoidal site to create two triangles.[27] At the midpoint of this line stands the column that bears his name. In this way, the architect's signature appears at the very heart and origin of the building.[28]

14.2
Building credits, East Building, National Gallery of Art (Washington DC), I.M. Pei and Partners

Like all of the museum buildings considered here, the East Building was lauded as an artwork in its own right; appropriately so, as Pei considers himself an artist.[29] Curiously, one level directly beneath Pei's signature is an unsigned wall drawing by Sol LeWitt.[30] For his wall drawings, LeWitt provided a 'plan' or diagram with a written set of instructions for others to execute, thereby separating conceptual invention from material execution. LeWitt worked for a time as a graphic designer in Pei's office, and the experience had a self-professed 'big effect' on his questioning of the relationship between modes of production and artistic authorship; he wrote: 'an architect doesn't go off with a shovel and dig his foundation and lay every brick. He's still an artist.'[31] LeWitt signs his name to the instruction certificate but not the completed wall drawings. While the architect Pei signed the results of his work, the artist LeWitt signed only the drawing for the work.

In designing the East Building, the planners looked to the original 1941 West Building by John Russell Pope who, like Flagg, had studied in Paris.[32] The West Building's large bronze tablet with the architects' names was studied in preparing the East Building's inscription.[33] In both cases, the architect's name is not alone, appearing with others involved in the creation of the building. The Pope building tablet is located in a small alcove behind Andrew Mellon's prominent donor memorial, because 'the tablet concerns [the donor] primarily, and those who carried out his ideas'.[34] Memos about the inscription's location reveal that the East Building's credits were deemed a 'secondary inscription' that was 'subsidiary' to the donors' names. While the presence of the architect's name along with the patron's can commemorate the patron's discernment and the value of their gift, an earlier location proposed for the building's credits was the end of a service corridor, beyond an elevator and coatroom.[35] Like building inscriptions throughout history, the architect is listed between the patron, above, and the builder, below. By reflecting societal hierarchy with the most powerful at the top, the incised order of the names also concretizes the movement of creation from idea to body. This descent from intent to thing follows the influential Platonic notion of creation in Western art.[36] The architect, in this system, mediates in between, Hermes-like (a statue of whom is in the centre of the West Building), by incarnating the idea, shuttling among efficient and material causes.

The East Building has solid marble stones in its knife-edge, 19° corner near the entrance that, for some reason, visitors ritualistically touch, imparting a dark stain to the marble at about eye-level that appeared within two months of the building's opening.[37] Director J. Carter Brown explained that 'people stroke the building like a pet'.[38] The same affection is shown for the inscription of Pei's name. Although surrounded by other, arguably more prestigious names, only Pei's has been touched by visitor's fingers, as they might a pristine or sacred object, so often as to be permanently outlined with a dark aura. Furthermore, the architecture firm's full inscription, 'I.M. Pei & Partners', does not receive equal attention, as primarily Pei's name shows the effect of repeated caresses. This curious action, made by a large number of people, suggests that the physical presence of the name stands as a token for the person. In traditional magic, a name is invoked, often by inscribing it on to a talisman, to have power over the person or thing named. Rome, like many cities, had a secret name to protect it from its enemies.[39] Leon Battista Alberti, architect and designer of his own alphabet for building inscriptions, mentions more than once the operational and psychological power of a name.[40] Perhaps, in an unselfconscious way, touching the inscription of Pei's name allows the visitor to participate in the seemingly magical act of incarnating an idea into the form of something as large as a building. The stain of fingers on the inscription gives the inert stone an aura and vivifies the power of the name. Although the architect acts through drawings to build at a distance, the inscribed signature maintains the authentic presence of the architect in the building, much as Walter Benjamin described the aura of the work of art.[41] The autograph as the indexical presence of the author is realized by the touch of the visitors to the museum.

Signature under partial erasure

> Hans Hollein
> 1972/82

Viennese architect Hans Hollein signed the Städtisches Museum Abteiberg (Mönchengladbach, 1982) (Figure 14.3).[42] As Hollein also has artworks in the collection, his signature appears both *on* and *in* the museum.[43] Because the museum entrance is at the top of a steeply sloping site, the diminutive entry pavilion is a sentinel on the roof plaza. A design sketch by Hollein shows the intent of descending through the floor of this sky platform into a dark 'sepulcher' of art.[44] Although the building lacks a façade, it has a signature.[45]

14.3
Architect's signature on entry stone, Städtisches Museum Abteiberg, Hans Hollein

The building signature is near the entry pavilion, on the upper left corner of a large, polished, white-marble slab. This signature is a mediated autograph – beginning by being written by the hand of the architect directly on the stone in black ink. The architect then directed the stonemason, despite protest to the contrary, to roughly chisel away the corner of the polished stone from around the signature.[46] In this way, the building is constructed beneath the architect's drawing. The building grows under his pen – claiming absolute authority for the architect.

The background, where the marble is chiseled around the signature, is finished with gold leaf. The gilt surface captures and reflects light, recalling heavenly inspired artistic genius. In Hollein's words: 'architecture – the expression of man himself – at once flesh and spirit [is] . . . a materialization of the spiritual.'[47] In the depth of the stone is revealed not darkness but shimmering light, making the architect an alchemist by transforming mute rock into architectural gold. Through the particular quality of his signature, Hollein expresses his relation to the building by writing his name on to the building under erasure.[48]

In Renaissance paintings, signatures participate in the play between the virtual reality of the image and the material reality of paint on canvas.[49] Hollein achieves a similar effect in his architecture through his refined use of materials and detailing, which creates a distance through paradox: the drawing both is and is not the building. Similarly, Richard Meier signed the Ara Pacis Museum (Rome, 2006) by sinking lead into his hand-written signature on stone over which runs a thin sheet of water.[50] The qualities of signatures tell the story of the architect's relation to the making of the building.

Signature buildings are not building signatures

Some of the most well-known, recent 'signature' art museums notably lack their architect's signature, including the Guggenheim Museum in Bilbao by Frank Gehry (1997), the Milwaukee Art Museum by Santiago Calatrava (2001) and the Contemporary Arts Center in Cincinnati by Zaha Hadid (2003).[51] Signature buildings are in fact quite different from signed buildings, as a signature building is a symbol that fixes its identity through its image, whereas the signed building is an index that expresses the presence of the architect.[52]

During an early twentieth-century debate on signing buildings, one architect wrote: 'the work itself is signature enough', expressing an opinion still widely held.[53] In sixteenth-century Venice, painters similarly often chose not to sign their work, because the growing vogue for collecting pictures developed the conceit that the 'connoisseur's eye and mind should recognize the quality of acquisitions rather than relying, quite literally, on a name'.[54] This attitude – that the work should 'speak for itself' – has unintended consequences, however. Since the artwork itself must be identifiable without a signature, it tends to restrict the architect's work to a signature image, frustrating future innovation. As Mark Wigley put it: '[the architect] may be fantasizing about doing a different kind of architecture but is afraid to threaten [his or her] known brand. If you were [. . .] unrestricted by the need to keep imitating yourself to maintain your signature, you might not do the kind of work that you are famous for.'[55] The paradox of a signature as the sign of originality that can be forged is paralleled by the paradox of a lack of a signature that compels the object to have an ossified signature appearance. Building signatures are not made to be recognized, but to be analysed.

Eugène-Emmanuel Violet-le-Duc, a major influence on Flagg and many others, was an enthusiastic advocate of signing buildings, in part by emphasizing medieval building signatures 'as if the artist were signing his painting or the author his book'.[56] He describes style, not as a trait of the finished object, but rather as residing in the act of design that is internally coherent with its intention and way of building. 'Style belongs to men and is independent of objects.'[57] In this sense, building signatures free the architect to invent a particular design with intrinsic relative qualities, rather than being narrowly constrained to provide a signature building as a fixed standard of quality as excellence.

Seemingly sympathetic with Viollet-le-Duc's idea of style, Peirce's writings on quality as firstness or 'self-containedness' emphasize that qualities are predominant in feeling and something 'peculiar and idiosyncratic'.[58] A quality need not be an actually realized sensation, so it doesn't inhere in an object, but simply is in itself as, for example, 'the dream of redness, saltiness, ache or grief'. In this way, Peirce's notion of quality is sympathetic with Viollet-le-Duc's idea of style: that it is potential and allows multitudinous manifestations. A signature building attempting to achieve some universal notion of quality as excellence lacks the 'eternal possibilities' of Peirce's notion of quality as an open work.[59] The architects' signatures, on the other hand, reveal quality as character and allow otherwise mute buildings, when signed, to speak to us.

Notes

1 Tim Anstey, 'Architecture and Rhetoric: Persuasion, Context, Action', in Tim Anstey, Katja Grillner and Rolf Hughes (eds), *Architecture and Authorship*, London: Black Dog, 2007, pp. 18–29.

2 Lucian, *Hist. Conscr.*, pp. 62–3. P.M. Fraser, *Ptolemaic Alexandria*, Oxford: Clarendon Press, 1972, Vol. 1, pp. 18f. W. Deonna, 'Sostratos de Cnide et la vertu des formules invisibles', *Revue Archéologique*, July–October, 1921, cinquieme serie, XIV, pp. 175–8.

3 Anthony Hughes, 'The Cave and the Stithy: Artists' Studios and Intellectual Property in Early Modern Europe', *Oxford Art Journal*, 1990, 13, pp. 34–48. Raphael sometimes signed his name to works that he designed but that were entirely executed by others. Rona Goffen, 'Raphael's Designer Labels: From the Virgin Mary to La Fornarina', *Artibus et Historiae*, 2003, vol. 24, no. 48, pp. 123–42, 124.

4 In Renaissance workshops, the master may have only painted the faces and hands of the primary figures, while subordinates usually executed the rest. Rona Goffen, 'Raphael', pp. 304, 319.

5 Charles Sanders Peirce, *Collected Papers of Charles Sanders Peirce*, 8 vols, vols. I–VI, 1931–5, Charles Hartshorne and Paul Weiss (eds), vols. VII–VIII, 1958, Arthur Burkes (ed.), Cambridge, MA: Harvard University Press, 1931–58. Claude Gandelman, 'The Semiotics of Signatures in Painting: A Peircian Analysis', *American Journal of Semiotics*, 1985, vol. 3, no. 3, p. 76.

6 Joanna Woods-Marsden, *Renaissance Self-portraiture: The Visual Construction of Identity and the Social Status of the Artist*, New Haven, CT: Yale University Press, 1998. Jules Guerin, 'The Mural Decorations Described by the Painter', *Art and Archaeology*, June, 1922, vol. 13, p. 259.

7 Karl Swoboda, *Peter Parler, Der Baukünstler und Bildhauer*, Wien: Verlag Anton Schroll, 1940. Corine Schleif, 'Nicodemus and Sculptors: Self-reflexivity in Works by Adam Kraft and Tilman Riemenschneider', *Art Bulletin*, 1993, vol. 75, pp. 599–626.

8 Gail Fenske, *The Skyscraper and the City: The Woolworth Building and the Making of Modern New York*, Chicago, IL: University of Chicago Press, 2008, p. 234. Charles Moore, who wrote his Ph.D. dissertation on water in architecture, takes the place of the Vitruvian lion scupper. Charles Moore, *Buildings and Projects, 1949–1986*, New York: Rizzoli, 1986, p. 186.

9 William Strickland, architect of the Tennessee State Capitol (1845–55), was interred in the basement wall following his wishes. Nell Mahoney, 'William Strickland and the Building of Tennessee's Capitol', *Tennessee Historical Quarterly*, June, 1945, vol. IV, no. 2, pp. 99–153. Christopher Wren's memorial tablet in St Paul's Cathedral speaks to the visitor: 'Reader, if you seek his monument, look around you': LECTOR, SI MONUMENTUM REQUIRIS, CIRCUMSPICE! William Chambers, *Treatise on the Decorative Part of Civil Architecture*, London: Lockwood, 1862, p. 263.

10 The inscription appears in duplicate symmetrically on the inside of the right spring block at both ends of the main arch on the Arco dei Gavi in Verona, built in the first century ACE. Theodorus Mommsen, *Inscriptionis Galliae Cisalpinae Latinae Corpus Inscriptionem Latinarum*, 5, Berolini, 1872, p. 3464. I thank Marco Frascari for his direction on this inscription. While Donderer believes these locations are due to the founder's grudging permission for the architect's signature, it may have a more symbolic reason as supporting the arch, and, as Donderer stresses, the patron need not have given permission at all. Michael Donderer, *Die Architekten der späten römischen Republik und der Kaiserzeit: Epigraphische Zeugnisse*, Erlangen: Universitätsbibliothek Erlangen-Nürnberg, 1996, p. 30. Modern historians have suggested that, as the Code of Justinian forbade architects to sign buildings, this Vitruvius may have been an Aedile, a government administrator, rather than the design or construction architect. Jefferson Pearse, 'The Organization of Roman Building During the Late Republic and Early Empire', Ph.D. dissertation, Trinity College, Cambridge, 1974. I. Calabi Limentani, *Studi Sulla Società Romana, Il Lavoro Artistico*, Milano: Istituto Editoriale Cisalpino, 1958, p. 89. The unlikely attribution of this signature to the Vitruvius who is the author of *de architectura* was debated in the Renaissance. Sebastiano Serlio, *Sebastiano Serlio on Architecture, Vol. 1, Books I–V, of 'Tutte l'Opere d'Architettura et Prospetiva'*, trans. Vaughan Hart and Peter Hicks, New Haven, CT: Yale University Press, 1996, p. 221. Among those inspired to sign their own work is Giovanni Maria Falconetto (1468–1535), also of Verona. Paola Marini, *Palladio e Verona: catalogo della mostra*, Verona: Pozza, 1980.

11 Gibbs, a student of Carlo Fontana in Rome (1704–8), also signed the Radcliffe Camera. Terry Friedman, *James Gibbs*, New Haven, CT: Yale University Press and Paul Mellon Centre for Studies in British Arts, 1984, p. 247.

12 Jennifer Alexander, 'Villard de Honnecourt and Masons' Marks', in Marie-Therese Zenner (ed.), *Villard's Legacy: Studies in Medieval Technology, Science and Art in Memory of Jean Gimpel*, Aldershot: Ashgate, 2004, pp. 53–69.

13 Pliny the Elder, *Natural History*, 36. 4. 42. Winckelmann identified the columns with those in the church of San Lorenzo outside the walls at Rome. Johann Joachim Winckelmann, *Anmerkungen über die Baukunst*, pp. 29–30, and *Monumenti Antichi ined.*, Vol. II, p. 269. It was common practice for Romans to represent names figuratively. Eugene McCartney, 'Canting Puns on Ancient Monuments', *American Journal of Archaeology*, 1919, 23, second series, pp. 59–64.

14 Early Christian-era churches in Syria have votive inscriptions with the architects' names beseeching divine aid. Michael Donderer, *Die Architekten*, p. 25. Catherine King, 'Effigies: Human and Divine', in Diana Norman (ed.), *Siena, Florence and Padua: Art, Society and Religion, 1280–1400*, Vol. 2, New Haven, CT: Yale University Press, 1995, pp. 124–7.

15 The marble façade is separated by a molded string course from the basement of rusticated granite sitting within a *fosse* or ditch that accommodates the building to its gently sloping site. Anonymous, 'The Works of Ernest Flagg', *Architectural Record*, April, 1902, 11, pp. 5–9, 6. Sven Sandström, *Levels of Unreality: Studies in Structure and Construction in Italian Mural Painting During the Renaissance*, Uppsala: Almqvist & Wiksell, 1963.

16 Rona Goffen, 'Raphael', p. 308. The Corcoran's attic story frieze carries the inscription: PHIDIAS, GIOTTO, DURER, MICHAELANGELO, RAPHAEL, VELASQUEZ, REMBRANDT, RUBENS, REYNOLDS, ALLSTON, INGRES. The artists were selected from a list of twelve compiled by Flagg, except for Dürer. One artist from the United States is included, Washington Allston, who was Flagg's uncle by marriage. Anonymous, *Architectural Record*, p. 6.

17 Mardges Bacon, *Ernest Flagg, Beaux-Arts Architect and Urban Reformer*, New York: Architecture History Foundation, 1986, p. 344, note 30.

18 Walter Muir Whitehill, 'The Making of an Architectural Masterpiece: The Boston Public Library', *American Art Journal*, Autumn, 1970, vol. 2, no. 2, pp. 13–35, 20–1.

19 Etching was taught at the École des Beaux-Arts, along with architecture, painting and sculpture. Ernest Flagg, 'The École des Beaux-Arts, First Paper', *Architectural Record*, January–March, 1894, vol. 3, pp. 302–13. There, the student of architecture 'is constantly brought in contact with the

greatest works of art in the other branches'. Ernest Flagg, 'The École des Beaux-Arts, Third Paper', *Architectural Record*, July–September, 1894, vol. 4, pp. 38–43. On the use of *invenit* on prints, see: David Rosand, *Drawing Acts: Studies in Graphic Expression and Representation*, Cambridge: Cambridge University Press, 2002, pp. 49, 352, note 41.

20 Russell L. Willard, 'Signed Buildings', *Pencil Points*, September, 1929, vol. 10, pp. 633–4. For example, Hector Guimard signed his buildings, including the Paris Metro stations. Maurice Rheims, *Hector Guimard*, New York: Abrams, 1985.

21 Joseph Rykwert, 'The Constitution of Bohemia', *RES* 1997, vol. 31, pp. 109–27. 'An architect in the true meaning of the term, that is to say an artist . . .' Ernest Flagg, 'The Justification of the Architect is Architecture', *American Architect and the Architectural Review*, March 29, 1922, vol. 121, p. 245. Julien Guadet, theory professor at the École, defines the architect as 'the artist who composes buildings'. Julien Guadet, *Éléments et Théorie de l'Architecture*, Paris, 1908, pp. 4, 504.

22 Ernest Flagg, 'Influence of the French School on Architecture in the United States', *Architectural Record*, October–December, 1894, vol. 14, p. 222. In addition to several portraits, Garnier signed the Opera 'CHARLES GARNIER ARCHITECT 1861–1875' in a lobby ceiling lunette at the heart of the building. Gérard Fontaine, *Charles Garnier's Opéra, Architecture and Interior Decor*, trans. Charles Penwarden, Paris: Centre des monuments nationaux, 2004, pp. 153, 170.

23 Sullivan also signed the numerous bank buildings he completed towards the end of his career.

24 Ernest Flagg, 'The Signing of Buildings', *The American Architect and Building News*, July 15, 1908, vol. 94, pp. 20–1. Flagg's signature is visible on the Charles Scribner Building (1913) 597 Fifth Avenue, New York City.

25 Joseph Rykwert, *The Judicious Eye, Architecture Against the Other Arts*, Chicago, IL: University of Chicago Press, 2008. Charles Platt, who studied painting at the École, signed the Freer Gallery (Washington, DC, 1923) within the context of this approach to art and architecture. Keith Morgan, *Charles A. Platt: The Artist as Architect*, Cambridge, MA: MIT Press, 1985.

26 The National Gallery of Art (Washington, DC) collection has numerous Renaissance paintings with signatures, including two portraits by Giovanni Bellini.

27 Michael Cannell, *I. M. Pei: Mandarin of Modernism*, New York: Carol Southern, 1995, p. 246.

28 The pier that holds the inscription enters the design only after numerous iterations, probably not until early in 1971. This pier was added to support an overhead mezzanine, to provide an airshaft for distributing fresh air, and to provide a location for the information desk. This is based on a review of the architect's design drawings in the National Gallery of Art Archives. Ground-floor plan drawing dated 21 July 1970, no. 403, does not show a column at the information desk, while a plan dated 28 May 1971 shows the column and studies its utility as an air supply. That Pei's name is inscribed on a stone airshaft also suggests that the breath of the building is related to the soul of the architect. I particularly thank Michelle Willens, Archivist, for her assistance in reviewing the materials held by the National Gallery Archives. See also the perspective drawing of 27 January 1971, in: I.M. Pei & Partners, *I. M. Pei & Partners Drawings for the East Building, National Gallery of Art, Its Evolution in Sketches, Renderings and Models, 1968–1978; An Exhibition Developed by Adams Davidson Galleries with the Cooperation of the National Gallery of Art and the American Institute of Architects Foundation*, Washington, DC: Adams Davidson Galleries, 1978.

29 'What I value most is that the East Building is a great work of art in its own right', J. Carter Brown, quoted in Andrea Oppenheimer Dean, 'The National East: An Evaluation of I.M. Pei's Landmark after Six Years', *Architecture: The AIA Journal*, October, 1984, vol. 73, no. 10, pp. 74–9, 78.

30 Sol LeWitt, Wall Drawing no. 681 C/A wall divided vertically into four equal squares separated and bordered by black bands. Within each square, bands in one of four directions, each with colour ink washes superimposed (1993). Sol LeWitt, *Wall Drawings 1984–1992*, Susanna Singer (ed.), Andover, MA: Philips Academy, 1992.

31 Sol LeWitt, 'Excerpts from a Correspondence, 1981–1983', in *Sol LeWitt, Wall Drawings 1968–1984*, Amsterdam: Stedelijk Museum, 1984, p. 23. Kirk Pillow, 'Did Goodman's Distinction Survive LeWitt', *Journal of Aesthetics and Art Criticism*, Autumn 2003, vol. 61, no. 4, pp. 365–80.

32 Steven Bedford, *John Russell Pope, Architect of Empire*, New York: Rizzoli, 1998.

33 'The question of the plaque was one of courtesy, at the expense of the Trust [. . .] Unless the Trustees were assured that all agreed, the trust will not install it', D. Shepard to Paul Mellon, 9 January 1941. There was a strong disagreement between Mrs Pope and the successor firm, Eggers and Higgins, which completed the building after Pope's death, over how to credit the

architects. The tablet as installed lists in order from the top: patron, building committee, architect, associate architects, construction supervisor, builder and workers. As if to complete the hierarchical ladder, earlier wording included thanking 'skilled craftsmen, mechanics and laborers', which, in the end, was simply shortened to 'many others who have contributed their excellent work'. Otto Eggers to Mellon Trust, November 28, 1940. Records of the A.W. Mellon Educational and Charitable Trust Construction Files. C. 1937–41, Series 12A1, Box 38, National Gallery of Art Archives (Washington, DC).

34 D.D. Shepard, Memo, July 23, 1941. Records of the A.W. Mellon Educational and Charitable Trust Construction Files. C. 1937–41, Series 12A1, Box 38, National Gallery of Art Archives (Washington, DC).

35 J. Carter Brown – Notes of Conversation with Mr. Mellon, Memo to David Scott, May 24, 1977. David Scott, Memo re East Building Inscriptions, March 14, 1977. National Gallery of Art, East Building, J. Carter Brown Office Files, Series 2C1, Box 15, National Gallery of Art Archives (Washington, DC).

36 Erwin Panofsky, *Idea, A Concept in Art Theory*, trans. Joseph Peake, Columbia, SC: University of South Carolina Press, 1968.

37 Michael Cannell, *I. M. Pei*, pp. 257, 274.

38 J. Carter Brown quoted in Andrea Oppenheimer Dean, 'National East', p. 75.

39 Henry Agrippa, *Three Books of Occult Philosophy*, trans. James Freake, Donald Tyson (ed.), St Paul, MN: Llewellyn, 1993, p. 213. Joseph Rykwert, *The Idea of a Town: The Anthropology of Urban Form in Rome, Italy and the Ancient World*, Princeton, NJ: Princeton University Press, 1976, p. 59.

40 Leon Battista Alberti, *De re aedificatoria* II, 13, VI, 4. Joseph Rykwert in Leon Battista Alberti, *The Ten Books of Architecture*, trans. James Leoni, New York: Dover, 1986 (1755), note 57. Christine Sperling, 'Leon Battista Alberti's Inscriptions on the Holy Sepulchre in the Cappella Rucellai, San Pancrazio, Florence', *Journal of the Warburg and Courtauld Institutes*, 1989, vol. 52, pp. 221–8.

41 Walter Benjamin, 'The Work of Art in the Age of Mechanical Reproduction', in Hannah Arendt (ed.), *Illuminations: Essays and Reflections*, New York: Schocken, 1968, pp. 217–52.

42 At the time of writing, the museum is closed and undergoing renovations. Rudi Fuchs, 'Municipal Museum at Abteiberg, Mönchengladbach, 1982', *Architecture and Urbanism*, January, 1984, vol. 160, no. 1, pp. 27–58.

43 Richard Morphet, 'The New Städisches Museum at Mönchengladbach', *The Burlington Magazine*, December, 1982, vol. 124, no. 957, pp. 759–61, 760. 'Hollein has signed the museum at one of its corners like a painter signing a canvas', Jonathan Glancey, 'Sterling and Hollein', *Architectural Review*, December, 1982, vol. 172, no. 1030, pp. 53–71, 64.

44 Kenchiku Toshi, 'Municipal Museum Abteiberg Mönchengladbach', *Architecture and Urbanism*, February, 1985, extra edition, vol. 2, pp. 66–91, 90. Miichi Miyake, 'Architecture: Beyond the Secret Objects', in *Abteiberg Museum, Mönchengladbach, West Germany, 1972–82, GA Document*, 1983, vol. 6, pp. 34–53, 34.

45 Anthony Vidler, 'Losing Face: Notes on the Modern Museum', *Assemblage*, June 1989, vol. 9, pp. 40–57.

46 Wolfgang Pehnt, *Hans Hollein Museum in Mönchengladbach, Architektur als Collage*, Frankfurt am Main: Fischer, 1986, p. 64. I thank Ulrike Altenmüller for her valuable translation.

47 Hans Hollein, 'Absolute Architecture', *Catalogue of the Hollein-Pichler Exhibition*, Vienna, 1963, in Ulrich Conrads, *Programs and Manifestoes on 20th-century Architecture*, Cambridge, MA: MIT Press, 1970, pp. 181–2.

48 For this argument in a different context, see: Joseph Koerner, *The Moment of Self-portraiture in German Renaissance Art*, Chicago, IL: University of Chicago Press, 1993, p. 423.

49 Rona Goffen, 'Raphael', pp. 313–7.

50 Paulette Singley, 'Roma Macchiata: The Stain of White', *Log*, Summer–Fall, 2007, vol. 10, pp. 129–36. I thank Renée Tobe for introducing this signature to me.

51 Zaha Hadid's Lois & Richard Rosenthal Center for Contemporary Art in Cincinnati (2003) is not signed, although her name is incised into the cardboard cover of the book devoted to her design, as if a substitute for its presence in the building.

52 But see: Wouter Davidts, 'Why Bother (About) Architecture? Contemporary Art, Architecture and the Museum', in Arie Graafland (ed.), *Crossover: Architecture, Urbanism*, Delft: Delft School of Design, 2006.

53 Frank Quinby, 'Committee Report on Signing Buildings and Using Institute Initials', *The Western Architect*, January, 1908, vol. 11, p. 4.

54 Louisa Matthews, 'The Painter's Presence: Signatures in Venetian Renaissance Pictures', *The Art Bulletin*, 1998, vol. 80, p. 641.

55 Mark Wigley, 'How Old is Young?', in *Famous: Perspecta 37, The Yale Architectural Journal*, Cambridge, MA: MIT Press, 2005, pp. 64–77, 67. See also: Peggy Deamer, 'Branding the Architectural Author', in *Famous: Perspecta 37*, pp. 44–5.

56 Viollet-le-Duc may have even included himself among the statuary in his restoration of Notre Dame. Eugène-Emmanuel Viollet-le-Duc, 'Architecte', in *Dictionnaire raisonné de l'architecture francaise du XIe au XVIe siècle*, Paris: Gründ, 1854–67, Vol. I, p. 111.

57 Eugène-Emmanuel Viollet-le-Duc, *The Foundations of Architecture, Selections from the Dictionnaire raisonné*, trans. Kenneth Whitehead, New York: Braziller, 1990, p. 233.

58 Charles Sanders Peirce, *Collected Papers of Charles Sanders Peirce*, vol. 1, Charles Hartshorne and Paul Weiss (eds), Cambridge, MA: Harvard, 1931, p. 302.

59 Peirce, *Collected Papers*, vol. 1, p. 426.

Chapter 15

Quality beyond measure

Architecture in the Lacanian account of culture

Catherine Belsey

Empty spaces

Architecture, as it happens, lies right at the heart of Jacques Lacan's theory of culture. The earliest architecture, he notes, 'can be defined as something organized around emptiness'.[1] This cryptic observation finally made sense to me when I first encountered Malta's prehistoric temples. The under-renowned and largely unaccountable presence of these remarkable stone structures would, I believe, have delighted the French psychoanalyst and confirmed his conviction that what gives us most pleasure in cultural artefacts is not subject to the kind of measurement dear to the quality controllers.

Nearly 6,000 years ago, the prehistoric inhabitants of the Maltese islands produced the first grand buildings we know of. Earlier than the pyramids and before Stonehenge, these extraordinary instances of Neolithic architecture were constructed out of dressed stone by a culture without writing. Very little is known of the people who built these magnificent edifices, still less of the motives that impelled a population of perhaps 10,000, who lived mainly by farming and fishing, to erect the thirty or more surviving structures now known, for want of a better word, as temples.[2]

Extant models indicate that this society supported architects. Since, in most cases, ancient buildings tend to conform to existing tradition, we are in the habit of assuming that they were not necessarily designed by identifiable individuals or groups of dedicated professionals. But in Malta the case is different. The handful of miniaturized versions found there by archaeologists are not just toys or decorative objects. Instead, elevations and cutaway terracotta ground plans of the temples, their walls and doorways represented to a low height, seem to be blueprints for the massive structures themselves.[3] Moreover, although there are strong resemblances among the temples, there are also distinctive new departures from one to another: innovation was evidently not alien to the culture that produced such a remarkable development over the course of the thousand years or so leading up to 2500 BC. The assumption must be that this stone-age community, living only just above subsistence level, supported skilled specialists who planned and supervised the construction work.

Approached by an open forecourt, each temple generally consisted of a tall, paved corridor, opening to apsidal spaces on each side and ahead. An imposing rectangular entrance at Mnajdra gives on to receding enclosures beyond (Figure 15.1). These chambers were originally roofed, often vaulted or domed. Though at this distance of time it is hard to be sure what practices they housed, the scale of the buildings is well in excess of anything required by any imaginable function. In the larger south temple at Ggantija, the most impressive of them all, surviving walls reach a height of seven metres in places, while the distance spanned by the inner apses is twenty-three metres from end to end. It seems that the grandeur of the areas the temples circumscribed must have signified in and of itself.

15.1
Entrance to the Lower Temple, Mnajdra, Malta

Six thousand years later, we live in a world preoccupied by quality control. It began with legislation to keep chalk out of flour and water out of beer, and progressed to the commercial supervision of cloth and classic sweaters. Marks and Spencer were leaders in quality control and have kept generations in underwear since securing their reputation in this field. But now the project has moved into areas where the measurement of quality is not so self-evidently appropriate, including education, as well as design and creativity. Are there criteria we can usefully bring to bear on these activities? How would we measure the quality of the Maltese temples? What would it mean to do so?

Jacques Lacan's Seminar of 1959–60 effectively produced a psychoanalytic theory of culture. When Lacan describes ancient architecture as something organized around emptiness, he is making the point that the first buildings carved out a space by putting a frame round it: where before was simply the visible, palpable world and its landscapes, now there were instances of structured absence. Buildings bring vacancy into existence, just as the work of the potter produces a hollow where before there was continuity. That a jar may be filled with liquid means it can also be emptied; a structure for habitation larger than is necessary purely for food and shelter creates spaces. Lofty chambers install a void.

Lacan distinguishes between use and meaning. He takes for granted the fact that a jar is a utensil and, by implication, that a building has a purpose and a function, but his theory of culture dwells on the signifying power of these objects. Pottery and architecture, crafts that go way back into the history of human culture, encircle, he proposes, a vacancy that alludes to a loss. He would call it the lost object in the inextricable real and would see it, moreover, as the ultimate cause of desire.

Confronted by my first Maltese temple, I found myself wondering what went on there. What ideas of divinity might this prehistoric society have held? Would their practices be intelligible to us as *religious* in our sense of the term? Archaeologists have formed opinions based on the material evidence, but they must in the last instance be conjectural. If the temples are a puzzle for us now, however, it seems their design inscribed another kind of mystery for the people who frequented them. A typical ground plan looks like a cloverleaf. Spaces opened beyond spaces, and, in some cases, further recesses beyond these. In the Hypogeum of Hal Saflieni, a labyrinthine underground construction of the same epoch, part temple, part necropolis, chambers give on to further chambers, whose walls in turn contain smaller cavities, their purpose now unknown. I am reminded indirectly of the evolution of medieval churches, where the simple rectangular ground plan soon gave way to a separation of the chancel, transepts opening on either side of the crossing, and then aisles and aisle-chapels, spaces once again giving on to further spaces in a kind of regress that marks, paradoxically, the progress of architectural skill.

In Malta, all the indications are that access to the inner chambers was restricted. Doors opening on to the interior apses were constructed to be closed from the inside. At Hagar Qim, where some of the main apses lead to additional bays, even the first chambers are accessible only by 'porthole' doorways, rectangles carved out of the centre of the megaliths themselves, and perhaps kept screened or veiled.[4] At the same time, in some places walls are pierced to allow the transmission of sound from the interior to the outer areas (Figure 15.2). What noises emerged from these so-called 'oracle holes'? Advice, commandments, singing, ceremonial chanting? That, from outside, the voices were effectively disembodied must have deepened the sense of the uncanny for their listeners.

Since there were no windows, these inner areas would also have been profoundly dark. There must presumably have been some form of artificial lighting, but no lamps have been found, though there are hearths. No lamps have survived at the underground Hypogeum either, however, where some of the walls still bear traces of red ochre painting. That at least testifies to an intended visibility, however intermittent. The experience of entering the temples from the open air would have meant moving from light into some degree, at least, of darkness, while perceiving or sensing spaces to which access was barred, glimpsing apertures that prohibited entry to the mysterious recesses beyond.

Enigma

Two thousand years after the collapse of the civilization that built the Maltese temples, Zeuxis made a fortune in Athens out of his artistic skills. The celebrated painter, as Lacan recounts, was able to depict grapes so accurately that birds came and pecked at them. But his younger rival, Parrhasius, outdid him in a competition when he painted a veil so convincing that Zeuxis asked him to reveal what lay behind it.

We smile knowingly at the naivety of Zeuxis. Obviously, we think, there was nothing behind the image of the veil: on the contrary, there was just paint on a material surface, probably wood. But there is no way to be sure of that. *Did* the veil conceal something? Had Parrhasius painted something behind it? Who can say

15.2
An internal doorway and 'oracle holes' at Mnajdra

for sure? One thing is clear, however: what is at stake in this contest is not just verisimilitude. That's only part of the story. The winning work is lifelike, but it also bears on uncertainty; it creates an enigma; and the implication must be that, in this instance at least, culture's most successful imitation is not of the world but of seduction. The work enlists the desire of the spectator and sustains it mimetically. By withholding its secret, the something or nothing it does *not* show, the painting of the veil mimics the ways of seduction and arouses desire.[5]

Where would the quality regimes have stood in this competition between Zeuxis and Parrhasius, I wonder? It might be possible to measure mimetic accuracy – at least, we could discuss it. But can seduction be measured, or can we quantify what might or might not be there? How would we manage or maximize enigma?

Behind the veil, Lacan would say, lies what he calls the *objet petit a* (object little a, but he didn't want the term translated). This object does not exist as such; it is, to all intents and purposes, empty. And for this reason, it stands in for the lost object that motivates desire. When Lacan rereads Sigmund Freud, he remodels Freud's account of desire as the impulse to regain the contentment of the infant in the arms of its mother. In Lacan's version, we lose organic oneness with the material world when language intrudes into that continuity, bringing difference and the possibility of absence. We learn language from outside ourselves, from other people, from our culture. Language releases human beings from submission to materiality by enabling us to name what is not present; it frees us to ask for what we want, or specify the place where we feel pain. But, at the same time, it shapes us by separating things out, identifying individual objects, introducing discontinuity between the little human being and the materiality that surrounds it. The language that defines us and our context bars re-entry into the gratification of a purely organic existence – always supposing there ever was such a condition.

For Lacan, as for Freud, civilization is, in consequence, a place of discontents. Many of our deepest endeavours are motivated by a desire to refind or replace what we seem to have lost. Freud thought sex was the best substitute; Lacan says sex merely 'occupies' the field of desire, as one sought-after gratification among others.[6] (Neither thought sex would entirely deliver.) Both agree that art, architecture, culture in its entirety, as well as science, are driven on ultimately by curiosity, restlessness, the will to know, an underlying, human dissatisfaction with what we already have; they concur that discontent is motivated in the last instance by the inaugural loss effected by the intrusion of culture into organic being.

Lacan returned again and again to the ways culture itself offers to compensate for that loss, or rather to placate the desire of both those who make it – architects, potters, artists – and also the rest of us – viewers, users or worshippers. In Lacan's account, the cultural objects that prove most seductive refer back to the initial loss of organic continuity. They hint at the emptiness that marks the place of what is missing, veiling it in such a way as to generate uncertainty. Is there something more beyond what we can see, they seem to ask, or is there nothing there? In other words, they keep us guessing.

Decoration

The Maltese temples at once draw attention to their own recesses and conceal them from ordinary view. Their darkness, doors, screens and oracle holes all point to what is *not* seen, inciting curiosity, generating uncertainty, enlisting the desire to know. At the same time, there was more to the Maltese temples than uncanny spaces, and there is more to Lacan's theory of culture than the representation of a loss that is the object-cause of desire. Why, he asks, do artefacts give pleasure? One striking feature of ancient buildings, as indeed of the pottery that constitutes their analogue in the enclosure of space, is the degree to which form and decoration take them beyond the realm of pure utility. From the earliest times, prehistoric Maltese ceramics were variously – and elegantly – shaped; they were polished, impressed or incised with fine lines. Bands or chevrons were inlaid in black or grey bowls, jars and vases. As time went on, the clay grew thinner, while the fashion changed to red, and then cream with red lines painted on it. At an early stage, meanwhile, rudimentary mimetic figures began to appear, scratched into the pottery. Schematic human shapes took the form of an X, with the addition of arms hanging straight down and a triangular head.

Colour, pattern, verisimilitude all offer to gratify the eye, delaying it in the restless quest for satisfaction. We are ultimately – in the most literal sense – looking for something. The painter, Lacan says, appeases the look in its quest, 'gives the eye something to feed on', and in the process invites the people to whom the picture is presented to lay down their gaze, 'as one lays down one's weapons. 'This', he goes on, 'is the pacifying, Apollonian effect of painting'.[7]

The Maltese temples, too, were lavishly decorated, especially their outer reaches, the places accessible to everyone. Stone surfaces were often pitted, the closely spaced indentations creating pattern and texture (Figure 15.3). Some walls were plastered, smoothing off the rough stone, and then painted with red ochre imported for the purpose. In the Hypogeum black and white chequers survive, as well as red ochre spirals. If not exactly a blaze of colour, the temples seem to have offered much to appease the eye. The main altar at Tarxien was elaborately carved with spirals, each itself further adorned with 'thorns' and 'fishtails'. The same temple boasted friezes showing remarkably lifelike animal figures – sheep and pigs among them.[8] That these may have had ritual significance does not detract from the pleasure they may also have provided to a society that seems to have lived in other respects at a fairly basic level.

What fascinated Zeuxis in ancient Athens was a painting executed with such skill that it aroused his desire to know what lay behind the veil it depicted. The point here is a double one: the paint itself, its colour, texture, fidelity to the world we recognize, offers to please the eye, but at the same time the image beckons beyond itself. There is nothing necessarily supernatural here, though Lacan's theory has been appropriated for quasi-religious purposes. His own interests were confined to the human condition. We are, he insisted, speaking beings, split between the participation in dialogue or other forms of cultural exchange that make us human, and the organisms we so evidently remain. The separation of the cultural subject from its own organic being generates a desire to refind a lost object that

15.3
Pitted stones, Mnajdra

seems to reside outside the limits of what culture permits us to know. Art, like pottery, like architecture, is capable of representing that divided state. The winning work presents a delightful surface even as it points to something else; in other words, it offers pleasure at the very moment that it also gestures beyond pleasure.

People have always decorated their interior walls, it seems. Cave paintings, like the posters in student flats, give the eye something to rest on. Medieval churches and Tudor mansions were polychrome inside to a degree that would offend the National Trust.[9] Periodic purges have proved ultimately ineffectual. Reformation extremists whitewashed over the Doom paintings, biblical narratives and Dances of Death designed to instruct and fascinate Catholic churchgoers, but they left the tracery and the vaulting, not to mention some of the stained glass. Meanwhile, inscriptions, painted monuments and wall tablets multiplied. Later, twentieth-century minimalism never did quite stick, except with a handful of purists: we cling to our pictures and photographs.

But it was an architect, Brunelleschi, who perfected the means by which art would open up the illusion of space *within* those pictures, creating in images made to adorn solid walls recesses that seem to pierce the enclosing walls themselves. Perspective painting, its receding lines converging on a vanishing point that is often veiled but intelligible within the picture itself, simulates on a flat surface the cavities hollowed out in the walls of the Maltese temples. In the process, it perfectly reproduces that combination of pleasure for the eye and allusion to the lost object that characterizes grand but decorated architecture.

Brunelleschi's first perspective painting depicted a building, the Baptistery in Florence. At this early stage, it was necessary to view the picture by means of a hole in the canvas through which the image appeared in a mirror. But the miracle of monocular perspective was rapidly made available without the accompanying apparatus, and by the end of the fifteenth century Leonardo da Vinci vastly expanded the refectory of Santa Maria delle Grazie in Milan when he depicted *The Last Supper* above the high table. Christ's head is positioned precisely at the vanishing point of this intensely architectural composition. In no time, domed ceilings were opening up to depictions of the Christian empyrean or the Greek pantheon, making heaven itself visible from inside the building to even the most earthbound of observers.

Architecture lent itself to perspective painting, since the straight lines of the roofs, steps and tiled courtyards depicted could be positioned to draw attention to the depth of field that geometry now made available. In Raphael's *School of Athens*, for example, the philosophers discuss and dispute in the foreground, while magnificent stone arches give on to and frame the sky, receding towards the absolute, which is understood to be the topic of the philosophical debates. To one side of the centre, Plato (in the likeness of Leonardo da Vinci) points upwards towards the transcendent Ideas, while, on the other, Aristotle's outstretched palm indicates the material world, which is his theme. Their respective foreheads just cross the lines that converge on the vanishing point. Sprawled on the steps, meanwhile, Diogenes the Cynic turns his back on such highfalutin nonsense. On the tiled ground, the mathematicians Pythagoras and Euclid bury their attention in their work. But all this is framed by the sense, which only a perspective architectural painting can give, of the infinity that lies beyond the best endeavours of human knowledge.

Architecture, then, plays a crucial part in the history of art. Meanwhile, Lacan himself makes a further, related point about the reciprocal influence of art on the history of architecture:

> From the moment when perspective was discovered in painting, a form of architecture appears that adopts the perspectivism of painting. Palladio's art, for example, makes this very obvious. Go and see Palladio's theater in Vicenza, a little masterpiece of its kind that is in any case instructive and exemplary. Neoclassical architecture submits itself to the laws of perspective, plays with them, and makes them its own. That is, it places them inside of something that was done in painting in order to find once again the emptiness of primitive architecture.[10]

I interpret this observation as a reference to a new way of presenting the building itself to the viewer. In place of the castle, or the medieval fortified palace, where the decorative surfaces look inwards onto courtyards and quadrangles, as they still do to a high degree in the colleges of Oxford and Cambridge, Palladio's symmetry, his classical porticoes and loggias are all there to be seen from the outside, from a distance, and they at once affirm and conceal the grandeur of the spaces within.

Judgement

I do not think Lacan's account of culture was designed to impose criteria of judgement. He himself was not much interested in assessing quality: he included in his list of objects that enclose vacancy a collection of matchboxes (empty, of course) and also macaroni. He was fascinated by the tricks art can play but that art critics are prone to despise: *trompe l'oeil*, for the illusion of depth it creates and then discredits;[11] and anamorphosis, because the something it constructs can, by a movement of the eye or a shift of position, be reduced to nothing again – or to a smudge.[12]

We could, however, if we put our minds to it, construct lists of preferred works in the light of his account. I offer some of mine, for what they are worth, while recognizing that it is all a matter of taste. Mine would include the short fictions of Jorge Luis Borges in their entirety, as well as *Gone with the Wind* for its sheer storytelling skill and its unresolved ending. I'd choose *Hamlet* (obviously): just what *is* his problem? And *Antigone*: *why* does she bury her brother, against Creon's regulations, knowing that the penalty is to be walled up alive in her own tomb? What is the unwritten, unalterable law she names that drives her: is it Justice? Kinship? Incest? Whatever her motive, Antigone is magnificent. I would certainly include the *Mona Lisa*, not just for the enigma of her smile, but for the precarious, improbable place she occupies, poised above a distant, magical landscape of rocks and lakes, misty by contrast with the sharp clarity of the figure itself.

The common ground here is an unresolved element. In all these cases, the work gives so much that corresponds to Lacan's decorated surfaces, but, at the same time, the characteristic that brings me back to it again and again is what it withholds, the something or nothing that would enable me to be done with it, at least for now. The mystery presented by all the instances I have listed tantalizes and seduces. I once went to an exhibition of Titian's paintings, where portrait after portrait, Venus after Venus fixed his or her melancholy gaze on a distant object off-screen and so invisible to the spectator. The gallery at that moment was full of desire. In a daze, I went round the show and then, without a moment's hesitation, joined the queue to see it all again.

My list of Lacanian works would include Chartres Cathedral and King's College Chapel, of course, as well as Rachel Whiteread's somehow desolate, somehow triumphant casts of empty spaces. When it comes to film, I would have to name *Casablanca*, inevitably, and perhaps *The Lives of Others* for its account of the strange, intense relationship between two men who barely meet in a world that has lost its humanity. Meanwhile, *The Third Man* shows the lost object perfectly symbolized by Harry Lime himself, until the camera finds him, when he takes up a new representative place as the figure who has replaced Vienna's lost civilization. What cannot be recovered is the partitioned city's former splendour, apparent in the faded grandeur of the buildings, some now reduced to rubble. Black and white photography and the skewed camera angles make visible the desolation of post-war Europe, figured as the damaged children we never see in a world now turned over to racketeers. The zither music, at first quintessentially Viennese, pleasurable, light-hearted, becomes more maddening with each later repetition. For a moment,

Holly Martins finds an instant of meaning in the sewer where he shoots his friend – with permission. But the moment is one of loss, not gain, and the final image of Holly alone among the falling leaves in the cemetery reaffirms the emptiness made palpable in a film where every shot seems perfectly executed.

The last thing I would want to do with this list is construct a canon of great works that everyone *ought to* admire, though some of my choices already have the status of cliché. I would expect a fairly high degree of cultural and historical relativity to play a part in anyone's selection, as well, of course, as personal preference. But whatever the differences between our judgements, anyone who went along with Lacan's account at all, who saw absence, or an allusion to the lost object of unconscious desire, as playing a part in the works they selected, would be talking about something almost impossible to quantify or judge. How could we manage or produce a quality that might *not* be there? I would guess it can't be done.

My fear, on the contrary, is that quality control is designed precisely to erase the element of uncertainty from what we do. We might not want to be kept guessing about our sweaters and underwear; we might prefer uniform standards of food purity and hygiene – though even there quality control threatens unpasteurized cheese, and uniformity has all but eliminated delicious varieties of tomatoes and apples. But, in the arts, there are no guarantees, except that predictability is likely to lead to mediocrity. This is not to say there are no standards – of accuracy, skill, materials. But the something else that, if Lacan has a point, might, paradoxically, be *missing*, is not amenable to assessment by quality regimes.

Notes

1. Jacques Lacan, *The Ethics of Psychoanalysis 1959–60 (Seminar 7)*, trans. Dennis Porter, London: Routledge, 1992, p. 135.
2. I am grateful to Ivan Callus, who introduced me to the temples and commented helpfully on an earlier draft of this paper.
3. David H. Trump, *Malta: Prehistory and Temples*, Malta: Midsea Books, 2004, pp. 81–2, 192–5, 215. Specific information about the temples is taken from this book and from *The Megalithic Temples of Malta*. Available at http://web.infinito.it/utenti/m/malta_mega_temples (accessed 28 May 2007).
4. This is suggested by perforations in some of the stones , see Trump, *Malta: Prehistory and Temples*, p. 76.
5. Jacques Lacan, *The Four Fundamental Concepts of Psycho-analysis*, trans. Alan Sheridan, London: Penguin, 1994, pp. 103–4, 111–12.
6. Jacques Lacan, *Écrits: A Selection*, trans. Alan Sheridan, London: Tavistock, 1977, p. 287.
7. Lacan, *Four Fundamental Concepts*, p. 101.
8. These are now in the National Museum of Archaeology in Valletta.
9. The National Trust is a charity devoted to the restoration and preservation of historic buildings and landscapes in England.
10. Lacan, *The Ethics of Psychoanalysis*, p. 136.
11. Lacan, *Four Fundamental Concepts*, p. 112.
12. Lacan, *The Ethics of Psychoanalysis*, pp. 135, 139 42; *Four Fundamental Concepts*, pp. 79–90.

Afterword

Utter *qualia*

Three probe/poems

Theodore Landrum

The suggestion that the slippery issue of quality in architecture may not be measurable by number points to a more fundamental problem of its being incommunicable in prose. This peculiar concern is cataclysmic for enthusiasts seeking lucid command of the medium of architecture! In lieu of commentary on this long-standing problem, I propose to offer its performance via incantational probe. What follows are three 'probe/poems', commenced as generative ground for the needed discourse on quality. These 'poems' attempt to fabricate quality as an arousal. As such, they aim to perform also as a sounding of those limits by which the throng in question – myself included – plays host to meaning. It is for this reason that they are called 'probe/poems'.

I thank the organisers of the Quality conference, the Welsh School of Architecture and the publisher of the present volume for their open-minded support of this discursive experiment. I would like to offer two further acknowledgements. First and foremost, these poems were written as part of an ongoing exchange with my partner, Lisa Landrum, whose dissertation on 'Architectural Acts' (which shows that architects mime and always have) *should* cause a minor revolution in the way architecture is taught. I also thank Arthur Schaller, Dean of the School of Architecture and Art at Norwich University (the one in Vermont), for his support.

Primary sources consulted include: Valery's 'Quality', Francis Ponge's *Pebble* and *Soap*, Poe's *Eureka*, Plato's *Timaeus*, Auden's *Prologue: The Birth of Architecture*, A.R. Ammons' *Sphere*, Gertrude Stein's 'An Acquaintance with Description', Charles Simic's 'Poetry and Experience', Vittorio Gregotti's essay 'Mimesis', and Wittgenstein's 'A Lecture on Ethics'.

A final note on the situation of 'performance': there were approximately fifteen conference delegates in the room, seated cautiously around the periphery. Before reading, I asked to have the projector turned off, the 'black-out' shade rolled up, and the door to the room opened. The light from a generous clerestory window animated the room nicely and brought palpable attention to those 'present'. I can only hope that my presentation of the following 'probe/poems' performed similarly.

HOW SPEAK so as to *stir*
so as it ...were
influential qualitative *winds* ?
how say so-as-to blow away
proud airs of fettered prudence ?

with what words
would one sway
uttermost-yoke and reins-reluctant to this
founding wish and WILL

for ever fond
and freshening

Well?

Let us FIND
and/or fix them NOT in large lucid languages
so clear of *wondering*
THIS!

Rather wRing
our vowed and levered CANT
(past jargon's lists) and hisss howl yawn kiss
fling far our worry wardy fare
and so fetch near this *foundling miraculousness*
and pause and breathe
these pleas in-the-making
and TAKE up
AWAKENING
with Mouths... Eyes... Hands... EARS...

OPEN WIDE

to *each* least *BECKONING'S* remarkable degree:

whhOOAw Hmnmnm PApai! papaiax!

OIMOI
euuuuu-eu EUIOU
O? O! OOOO aaaaaahh

HA! eh ?

O a R swear momentous

hEUREKAStyx! in the intercoastal tub and liquid reckoning of *open* minds'
mosaic depths

OR e risk inventing:

hHh! conflating inwyrd sup with ‘magination's

resuscitating KIND of surprising inspyrelation

Ur Goin' *bach* to baby lib Taliesin's
immortal muting charmmmm:

blerwm blerwm blerwm blerwm blerwm
blerwm blerwm blerwm blerwm

and ‘otherwise’ sway by placing
moot
our amble queries
sounding schemas
probing stances

in the manner of a *pensive person* composing motley MOTIVES

How else host well-meaning's wink

to kindle *Qualia* and MAKE

unsink

Our tangled *gist* and *spark!*

Our SUSPECT SITUATION ?

LANGUAGE & ARCHITECTURE (*A DUOLOGUE*)

A ct Articulate Avoid & Annoy
B ustle Burgle Bully Browse & Brag
C orrect Calm & Coax
D isguise Dwindle & Delight
Architecture & Language
E mbed Epitomize Estrange & Exhume
F igure Flummox Flatter Fork & Fix
G esticulate Glamorize Gamble Glom & Gerrymander
Language & Architecture
H allow Herd Hide & Heed
In volve Ignite Idolize & Ignore
J ubilate Jinx Juggle & Join
K indle Keep Kiss & Kill
Architecture & Language
L oom Lobby Lather & Liken
M ine Mend Mourn Mystify & Meander
N ormalize Nationalize Nudge & Need
Language & Architecture
O fficiate Opacify Ostracize & Overdo
P ersist Placate Perplex & Probe
Q uarry Quilt Quibble & Queue
R ouse Rehearse Rank Ramble & Reveal
S leep Slur Stipulate Situate & Surmise
Architecture & Language
T ickle Topple Temper & Transform
U rge Unite Usher Understate & Undo
V acillate Veil Vex & Vow
Language & Architecture
W affle Wizen Wake Whip Warp & Woof
X-rate & Xeno-philanthropize
Architecture & Language
Y ield Yoke Yearn & Yawn
Z ip Zonk Zenith & Zero

INAUGURIES

. . . as before
or again and again
primordial latent *qualia*
gurgle in the potent bowl of *populus*
rather tracing tickle forth
from concave cusp and crave
as vexing mists
froth forth from rippled nightly surge
all else dormant curiosity
save these rare anticipant dribs
willed animas gather round
qua expectant earthen earhooks
lapping and loosing scrap sound shares
as whole as kindled kind can fête
peculiar portioned feats
simply put:
as phanta-plasmic organ-o-flits
fit about worlds of interest
merely made articulable
ephemeral generations exhaust themselves
on breach of memory
wild popular ever flattening gob
gone fundamental fragmentary same
and again, as before
early fabulous features
rolling boiling already burble over
under spheral plenum's porous crock
percolating in the effervescent flitch and swale
of carbonated mucky-muck ripe with perturbation
fecund as permeable to flim-flam flocculus
and sample candy fathomed core
as essence treat
as splice and store
of capability's ample steepening
simply put:
before one is in or on to the such and such degree
one has yet to put upon
persnickety thresh of casual devising

but *this*
as soon is soon enough
so as to stress
what chastening storm it is that is
and so behind beneath its building
nevertheless
plausibilities certain and yet so weird
forthwith
rhythmic haste must on and on
get on in with IF
wherein thereafter hence thence and whence
whether Will can find its nipple mouth
and know not why we root about
and so admit by often seeming so content
which is to say
surprising as it is
enough to field spontaneous 'arriving'
as turtled anticipation defends vague finding:
rehearse transmissive potentencies and linger
in the nested test of toiling temper-tude
as ever found best what most least tempted
our anytime's public pending seedling gone genera
vaulting knot-laced traps of mutual premonition
back against the storied nut of wall
as vital fade and tidal verge
on harboured feelings' splendid draw and splurge
on fancy taken in a curb
moreover
clarity's back door knock schemata
need pit against
one's see-through palace of fit occasions
emerging thresholds of buoyant pressing bounty
here too we now merge on this
qualified ballistic spore criterion
(as far as willing interests
invest at all in the fulcrum arts of causal being
or gravitate
to keenest tilting kind of ludic disposition)
and again, as before
advance a figure
mystery enhancing
as furthering as the mingle hastening crowd
profuse peculiar groundling spoil
fortuitous thwart and foil
as intractable as danger to the free

Select bibliography

Adcock, Craig, *James Turrell: The Art of Light and Space*, Berkeley, CA: University of California Press, 1990.

Ahlberg, Hakon, *Sigurd Lewerentz 1885–1975: The Dilemma of Classicism*, London: Architectural Association, 1989.

Aida, Takefumi, *Takefumi Aida: Buildings and Projects*, New York: Princeton Architectural Press, 1990.

Alexander, Christopher, *Notes on the Synthesis of Form*, Cambridge, MA: Harvard University Press, 1964.

Alberti, Leon Battista, *The Ten Books of Architecture*, trans. James Leoni, New York: Dover, 1986.

Amery, Colin, *Architecture, Industry and Innovation: The Early Works of Nicholas Grimshaw and Partners*, London: Phaidon, 1995.

Andrew, David S., *Louis Sullivan and the Polemics of Modern Architecture: The Present Against the Past*, Chicago, IL: University of Illinois Press, 1986.

Andrews, Edward Deming, *The People Called Shakers*, New York: Dover Publications, 1953.

Anstey, Tim, Grillner, Katja and Hughes, Rolf (eds), *Architecture and Authorship*, London: Black Dog, 2007.

Anthony, Kathryn, *Design Juries on Trial: The Renaissance of the Design Studio*, New York: Van Nostrand Reinhold, 1991.

Architects Journal, 'Guide to the *AJ* Information Library', *AJ*, 11 October 1972, vol. 41, no. 156, pp. 833–48.

Atkinson, George, *Guide Through Quality Control Standards*, Wokingham: Van Nostrand Reinhold, 1986.

Baden-Powell, Charlotte, Hetreed, Jonathan and Ross, Anne, *Architect's Pocket Book*, Oxford: Architectural Press, 2008.

Banham, Reyner, 'History Faculty, Cambridge', *Architectural Review*, December 1968, vol. 144, p. 329.

Barton, Nicholas, *The Lost Rivers of London*, London: Historical Publications, 2000.

Beardsley, John, *Earthworks and Beyond: Contemporary Art in the Landscape*, New York: Abbeville Press, 1984.

Bedford, Steven, *John Russell Pope, Architect of Empire*, New York: Rizzoli, 1998.

Benjamin, Walter, 'The Work of Art in the Age of Mechanical Reproduction', in Hannah Arendt (ed.), *Illuminations: Essays and Reflections*, New York: Schocken, 1968.

Benson, Ciaran, *The Cultural Psychology of Self: Place, Morality and Art in Human Worlds*, London: Routledge, 2001.

Beyer, Andreas and Boucher, Bruce (eds), *Piero de' Medici 'il Gottoso': Art in the Service of the Medici*, Berlin: Akademie Verlag, 1993.

Blundell Jones, Peter, 'Sigurd Lewerentz: Church of St Peter, Klippan, 1963–66', *Architectural Research Quarterly*, 2002, vol. 6, pp. 159–73.

Bourdieu, Pierre, *Distinction: A Social Critique of the Judgement of Taste*, trans. Richard Nice, Cambridge, MA: Harvard, 1984.

Bourdieu, Pierre, *Homo Academicus*, Stanford, CA: Stanford University Press, 1990.

Bourdieu, Pierre, *The Field of Cultural Production*, Cambridge: Polity Press, 1993.

Bowyer, Jack, *Practical Specification Writing: A Guide for Architects and Surveyors*, London: Hutchinson, 1981.

Boyarsky, Alvin, 'Stirling "Dimostrationi" ', *Architectural Design*, October 1968, vol. 38.

Bradley, John Lewis, *Ruskin's Letters from Venice 1851–1852*, New Haven, CT: Yale University Press, 1955.

British Standards Institute, *British Standards BS 4778–2:1991, Quality Vocabulary – Part 2: Quality Concepts and Related Definitions*, London: British Standards Institute, 1991.

British Standards Institute, *British Standards BS 5606:1990 Guide to Accuracy in Building*, London: British Standards Institute, 1990.

British Standards Institute, *British Standards BS 5628–3:2005 Code of Practice for the Use of Masonry – Part 3: Materials and Components, Design and Workmanship*, London: British Standards Institute, 2005.

British Standards Institute, *British Standards BS ISO 9000–2:1997 Quality Management and Quality Assurance Standards – Part 2: Generic Guidelines for the Application of ISO 9001, ISO 9002 and ISO 9003*, London: British Standards Institute, 1997.

Brown, Julia (ed.), *Occluded Front: James Turrell*, Los Angeles, CA: Fellows of Contemporary Art and the Museum of Contemporary Art, and Larkspur Landing, CA: The Lapis Press, 1985.

Bynum, W.F. and Porter, Roy (eds), *Medicine and the Five Senses*, Cambridge: Cambridge University Press, 1993.

Caldenby, Claes, Caruso, Adam and Lind, Sven Ivar, *Sigurd Lewerentz: Two Churches*, Stockholm: Arkitektur Förlag AB, 1997.

Cannell, Michael, *I.M. Pei: Mandarin of Modernism*, New York: Carol Southern, 1995.

Carolin, Peter and Dannatt, Trevor (eds), *Architecture, Education and Research: The Work of Leslie Martin – Papers and Selected Articles*, London: Academy Editions, 1996.

Cenicacelaya, Javier, 'La Iglesia De San Pedro En Klippan: La Belleza De Lo Directo (The Church of St Peter in Klippan: The Beauty of Directness)', *Via arquitectura*, 2004, vol. 14, pp. 8–13.

Clark, Thomas D. and Ham, F. Gerald, *Pleasant Hill and Its Shakers*, Harrodsburg, KY: Pleasant Hill Press, 1987.

Cohen, Matthew A., 'Ugly Little Angels: Deliberately Uneven Construction Quality in the Basilica of San Lorenzo in Florence', *Architectural Research Quarterly (arq)*, December 2007, vol. 11, no. 3/4, pp. 277–89.

Cohen, Matthew A., 'How Much Brunelleschi? A Late Medieval Proportional System in the Basilica of San Lorenzo in Florence', *Journal of the Society of Architectural Historians*, March 2008, vol. 67, p. 1.

Conti, Piero Ginori, *La Basilica di S. Lorenzo di Firenze e la famiglia Ginori*, Florence: Fondazione Ginori Conti, 1940.

Cook, E.T., *The Life of John Ruskin, Vol. 2 1860–1900*, London: George Allen, 1912.

Cook, E.T. and Wedderburn, Alexander (eds), *Works of John Ruskin*, London: George Allen, 1904–13.

Coyne, Richard and Snodgrass, Anthony, *Interpretation in Architecture*, London and New York: Routledge, 2006.

Crinson, Mark and Lubbock, Jules, *Architecture: Art or Profession? Three Hundred Years of Architectural Education in Britain*, Manchester: Manchester University Press, 1994.

Dal Co, Francesco, *Tadao Ando*, London: Phaidon, 1995.

Damasio, Antonio, *The Feeling of What Happens: Body, Emotion and the Making of Consciousness*, London: Vintage, 2000.

Dannatt, Trevor, 'Trevor Dannatt Praises the Profound Love of Quality', *RIBA Journal*, August 1973, pp. 382–3.

Davidts, Wouter, 'Why Bother (About) Architecture? Contemporary Art, Architecture and the Museum', in Arie Graafland (ed.), *Crossover: Architecture, Urbanism*, Delft: Delft School of Design, 2006.

de Maria, Walter, 'Some Facts, Notes, Data, Information, Statistics and Statements', *Artforum*, April 1980.

de Roover, Raymond, *The Rise and Decline of the Medici Bank: 1397–1494*, New York: W.W. Norton, 1966.

Dennett, Daniel, *Consciousness Explained*, London: Penguin, 1993.

Department of Trade and Industry, *Rethinking Construction*, London: HMSO, 1998.

Dewluf, G.and Van Meel, J., 'Sense and Nonsense in Measuring Design Quality', *Building Research and Information*, 2004, vol. 32, no. 3, pp. 247–50.

Donderer, Michael, *Die Architekten der späten römischen Republik und der Kaiserzeit: Epigraphische Zeugnisse*, Erlangen: Universitätsbibliothek Erlangen-Nürnberg, 1996.

Donnelly, Mark P., *Sixties Britain: Culture, Society and Politics*, Harlow: Longman, 2005.

Dunn, Lee, *The Student Assessment Handbook*, London: Routledge Falmer, 2004.

Durand, J.N.L., *Recueil et parallele des edifices*, Paris: n.p., 1801.

Edman, Bengt, 'Lewerentz the Bricklayer', *Spazio E Societa*, 1993, vol. 16, no. 64, pp. 76–83.

Eisner, Elliot W., *The Art of Educational Evaluation*, London and Philadelphia: Falmer, 1985.

Fenske, Gail, *The Skyscraper and the City: The Woolworth Building and the Making of Modern New York*, Chicago, IL: University of Chicago Press, 2008.

Fiennes, Celia, *The Journeys of Celia Fiennes*, Christopher Morris (ed.), London: The Cresset Press, 1947.

Fiennes, Celia, *The Illustrated Journeys of Celia Fiennes 1685–c1712*, Christopher Morris (ed.), London: Macdonald, 1982.

Filler, Martin, 'Santiago Calatrava', in *Makers of Modern Architecture: From Frank Lloyd Wright to Frank Gehry*, New York: New York Review of Books, 2007.

Flagg, Ernest, 'The Signing of Buildings', *The American Architect and Building News*, July 15, 1908, vol. 94, pp. 20–1.

Flora, Nicola, Giardello, Paolo and Postiglione, Gennaro (eds), *Sigurd Lewerentz 1885–1975*, Milan: Electa Architecture, 2001.

Floyer, Sir John, *An Enquiry Into the Right Use and Abuses of the Hot, Cold, and Temperate Baths in England, etc.*, London: R. Clavel, 1697.

Forty, Adrian, *Words and Buildings*, London: Thames & Hudson, 2001.

Frampton, Kenneth (ed.), *A New Wave of Japanese Architecture*, New York: Institute for Architecture and Urban Studies, 1978.

Fraser, P.M., *Ptolemaic Alexandria*, Oxford: Clarendon Press, 1972.

Freeman, Henry William, *The Thermal Baths of Bath: Their History, Literature, Medical and Surgical Uses and Effects; Together with the Aix Massage and Natural Vapour Treatment* (S.l.), London: Hamilton, Adams, 1888.

Friedman, Terry, *James Gibbs*, New Haven, CT: Yale University Press and Paul Mellon Centre for Studies in British Arts, 1984.

Gandelman, Claude, 'The Semiotics of Signatures in Painting: A Peircian Analysis', *American Journal of Semiotics*, 1985, vol. 3, no. 3, p. 76.

Gann, D.M., Salter, A.J. and Whyte, J.K 'The Design Quality Indicator as a Tool for Thinking', *Building Research and Information*, vol. 31, no. 5, pp. 318–33.

Gann, D.M. and Whyte, J.K., *Performance Indicators for the Design of Buildings, Final Report*, Ref. No. CIC00003, London: DETR, 2003.

Gibbs, D.D., 'Floyer, Sir John (1649–1734)', *Oxford Dictionary of National Biography*, Oxford: Oxford University Press, 2004.

Girouard, Mark, *Big Jim: The Life and Work of James Stirling*, London: Chatto & Windus, 1998.

Glancey, Jonathan, 'Sterling and Hollein', *Architectural Review*, December, 1982, vol. 172, no. 1030, pp. 53–71.

Gowing, Lawrence (ed.), *The Critical Writings of Adrian Stokes, Vol. 1 1930–1937; Vol. 2 1937–1958; Vol. 3 1955–1967*, London: Thames & Hudson, 1978.

Guidott, Thomas, *An Apology for the Bath, Being an Answer to a Late Enquiry (by Sir J. Floyer), Into the Right Use and Abuses of the Baths in England*, London, 1705.

Hall, Francis, 'Specifying for Quality', *The Architects Journal*, 8 June 1994, vol. 199, p. 23.

Hancock, David, *Master Builders of the Animal World*, New York: Harper & Row, 1973.

Hawkes, Dean, 'Sustainability and Architectural Synthesis', in *Architecture City Environment: Proceedings of PLEA 2000, Cambridge, UK, 2–5 July 2000*, London: James & James, 2000.
Hawkes, Dean, *The Environmental Tradition: Studies in the Architecture of Environment*, London: E&FN Spon, 1995.
Hayashiya, Tatsusaburo, Nakamura, Masao and Hayashiya, Seizo, *Japanese Arts and the Tea Ceremony*, trans. Joseph P. Macadam, New York: Weatherhill, 1974.
Hays, K. Michael, *Architecture Theory Since 1968*, Cambridge, MA: MIT Press, 1999.
Herbert, Lynn M. *et al.*, *James Turrell: Spirit and Light*, Houston, TX: Contemporary Arts Museum, 1998.
Hillier, Jean and Rooksby, Emma (eds), *Habitus: A Sense of Place*, Ashworth: Aldershot, 2002.
Hilton, Tim, *John Ruskin. The Early Years 1819–1859*, New Haven, CT and London: Yale University Press, 1985.
Hughes, Anthony, 'The Cave and the Stithy: Artists' Studios and Intellectual Property in Early Modern Europe', *Oxford Art Journal*, 1990, vol. 13, pp. 34–48.
Humphrey, Nicholas, *Seeing Red, A Study in Consciousness*, Cambridge, MA: The Belknapp Press of Harvard University Press, 2006.
Hyman, Isabelle, 'Notes and Speculations on S. Lorenzo, Palazzo Medici, and an Urban Project by Brunelleschi', *JSAH*, May 1975, vol. 34, no. 2.
Hymers, Paul, *Home Conversion*, London: New Holland, 2003.
Isozaki, Arata, *Ma: Space-Time in Japan*, New York: Cooper-Hewitt Museum, 1979.
Jencks, Charles, *Modern Movements in Architecture*, London: Penguin Books, 1982.
Jodidio, Philip, *Tadao Ando: Complete Works*, Cologne: Taschen, 2004.
Kent, Dale, *Cosimo de' Medici and the Florentine Renaissance*, New Haven, CT and London: Yale University Press, 2000.
Kite, Stephen, *Adrian Stokes. An Architectonic Eye*, London: Legenda, 2009.
Kite, Stephen, 'The Urban Landscape of Hyde Park: Adrian Stokes, Conrad and the *topos* of Negation', *Art History*, June 2000, vol. 23, p. 2.
Klingmann, Anna, *Brandscapes: Architecture in the Experience Economy*, Cambridge, MA: MIT Press, 2007.
Kubler, George, *The Religious Architecture of New Mexico: In the Colonial Period and Since the American Occupation*, Albuquerque, NM: University of New Mexico Press, 1990.
Lacan, Jacques, *Écrits: A Selection*, trans. Alan Sheridan, London: Tavistock, 1977.
Lacan, Jacques, *The Ethics of Psychoanalysis 1959–60 (Seminar 7)*, trans. Dennis Porter, London: Routledge, 1992.
Lacan, Jacques, *The Four Fundamental Concepts of Psycho-analysis*, trans. Alan Sheridan, London: Penguin, 1994.
Lakoff, Georg and Johnson, Mark, *Metaphors We Live By*, Chicago, IL: University of Chicago Press, 2003.
Lavin, Irving, 'Donatello's Bronze Pulpits in San Lorenzo and the Early Christian revival', in *Past–Present*, Berkeley, CA. University of California Press, 1993.
Leiman, Kirsi, *Concrete Spaces: Architect Aarno Ruusuvuori's Works from the 1960s*, Helsinki: Museum of Finnish Architecture, 2000.
Le Corbusier, *The Decorative Art of Today*, London: The Architectural Press, 1987.
LeWitt, Sol, *Wall Drawings 1984–1992*, Susanna Singer (ed.), Andover, MA: Philips Academy, 1992.
Littlefield, David (ed.), *Metric Handbook: Planning and Design Data*, Amsterdam: Elsevier/Architectural Press, 2008.
Littleton, C. Scott, *Understanding Shinto*, London: Duncan Baird, 2002.
Lloyd-Thomas, Katie (ed.), *Material Matters: Architecture and Material Practice*, London: Routledge, 2007.
Macmillan, Sebastian (ed.), *Designing Better Buildings: Quality and Value in the Built Environment*, London and New York: Spon, 2004.
Mallgrave, Harry Francis and Ikonomou, Eleftherios (eds), *Empathy, Form and Space: Problems in German Aesthetics, 1873–1883*, Santa Monica, CA: Getty Center Publications, 1994.

Manetti, Antonio, *Vita di Filippo Brunelleschi*, Giuliano Tanturli (ed.), Milan: Edizioni il Polifilo, 1976.

March, Lionel and Steadman, Philip, *The Geometry of Environment: An Introduction to Spatial Organisation in Design*, London: RIBA, 1971.

Markus, T.A., 'Lessons from the Design Quality Indicator', *Building Research and Information*, 2003, vol. 31, no. 5, pp. 399–405.

Marsh, David and Rhodes, R.A.W., *Implementing Thatcherite Policies: Audit of an Era*, Buckingham and Philadelphia, PA: Open University Press, 1992.

Martin, Leslie, 'Conference on Architectural Education', *RIBA Journal*, 1958, vol. 6, pp. 279–82.

Martin, Leslie, 'Education around Architecture: Address to the RIBA Conference, Birmingham, 1970', *RIBA Journal*, September 1970.

Martin, Leslie, *Buildings and Ideas 1933–83: From the Studio of Leslie Martin*, Cambridge: Cambridge University Press, 1983.

Martínez, Antonio Riggen, *Luis Barragán: Mexico's Modern Master, 1902–1988*, New York: Monacelli Press, 1996.

McKean, John, *Royal Festival Hall: London County Council, Leslie Martin and Peter Moro*, London: Phaidon, 1992.

McKean, John, *Leicester University Engineering Building: James Stirling and James Gowan*, London: Phaidon, 1994.

Medawar, P.B., *The Art of the Soluble: Creativity and Originality in Science*, Harmondsworth: Penguin, 1969.

Melcher, Marguerite Fellows, *The Shaker Adventure*, Cleveland, OH: The Press of the Case Western Reserve University, 1968.

Moore, Charles, *Buildings and Projects, 1949–1986*, New York: Rizzoli, 1986.

Moore, Rowan (ed.), *Structure, Space and Skin: The Work of Nicholas Grimshaw & Partners*, London: Phaidon, 1993.

Morgan, Keith, *Charles A. Platt: The Artist as Architect*, Cambridge, MA: MIT Press, 1985.

Morolli, Gabriele and Ruschi, Pietro (eds), *San Lorenzo, 393–1993, l'architettura: le vicende della fabbrica*, Florence: Alinea, 1993.

Mostafavi, Mohsen and Leatherbarrow, David , *On Weathering: The Life of Buildings in Time*, Cambridge, MA: MIT Press, 1993.

Naito, Akira and Nishikawa, Takeshi, *Katsura: A Princely Retreat*, Tokyo: Kodansha, 1994.

National Building Specification 1973, 1st edition, London: NBS Ltd, 1972.

Neufert, Ernst, *Architectural Data*, Oxford: Blackwell Science, 1995.

Nietzsche, Frederich, *The Nietzsche Reader*, ed. and trans. R.J. Hollingdale (1881), Harmondsworth: Penguin, 1977.

Noever, Peter (ed.), *James Turrell: The Other Horizon*, Ostfildern-Ruit: Cantz Verlag, 1999.

Norri, Marja-Riitta and Kärkkäinen, Maija, *Aarno Ruusuvuori: Structure is the Key to Beauty*, Helsinki: Museum of Finnish Architecture, 1992.

Odgers, Juliet, Samuel, Flora and Sharr, Adam (eds), *Primitive: Original Matters in Architecture*, London: Routledge, 2006.

Okakura, Kakuzo, *The Book of Tea* (1906), reprint, Rutland, VT: Charles E. Tuttle, 1956.

O'Malley, Therese and Treib, Marc, (eds), *Regional Garden Design in the United States*, Washington, DC: Dumbarton Oaks, 1995.

Ono, Sokyo, *Shinto: The Kami Way*, Rutland, VT: Charles E. Tuttle, 1962.

Ostler, Timothy, 'Does Quality Assurance Work?', *The Architects Journal*, 16 February 1994.

Pallasmaa, Juhani, *Eyes of the Skin: Architecture and the Senses*, London: Academy Editions, 1996.

Panciroli, Virginia (ed.), *Katsura: Imperial Villa*, Milan: Electa Architecture, 2005.

Pearman, Hugh, *Equilibrium: The Work of Nicholas Grimshaw and Partners*, London: Phaidon, 2000.

Pehnt, Wolfgang, *Hans Hollein Museum in Mönchengladbach, Architektur als Collage*, Frankfurt am Main: Fischer, 1986.

Peirce, Charles Sanders, *Collected Papers of Charles Sanders Peirce*, Charles Hartshorne and Paul Weiss (eds), Cambridge, MA: Harvard, 1931.

Pelkonen, Eeva-Liisa and Albrecht, Donald, *Eero Saarinen: Shaping the Future*, New Haven, CT: Yale University Press, 2006.

Penny, Nicholas, *Ruskin's Drawings*, Oxford: Phaidon, 1989.

Popper, Karl, *Objective Knowledge: An Evolutionary Approach*, Oxford: Clarendon Press, 1972.

Powell, C.G., *The British Building Industry Since 1800: An Economic History*, London: E&FN Spon, 1996.

Prasad, Sunand 'Clarifying Intentions: The Design Quality Indicator', *Building Research and Information*, Nov.–Dec. 2004, vol. 32, no. 6, pp. 548–51.

Rice, Ian, 'Ziggurats for Bureaucrats': Sir Leslie Martin's Whitehall Plan, *arq*, 2004, vol. 8, no. 3–4, pp. 313–23.

Rispa, Raúl (ed.), *Barragán: The Complete Works*, New York: Princeton Architectural Press, 1996.

Rolls, Roger, *The Hospital of the Nation: The Story of Spa Medicine and the Mineral Water Hospital at Bath*, Bath: Bird, 1988.

Royal Institute of British Architects Plan of Work, 1st edition, London: RIBA, 1967.

Rubinstein, Nicolai, *The Government of Florence Under the Medici (1434 to 1494)*, 2nd edition, Oxford: Oxford University Press, 1997.

Ruskin, John, *The Stones of Venice*, edited and abridged by J.G. Links, London: Pallas Athene in conjunction with Ostara Publishing, 2001.

Rykwert, Joseph, 'Florey Building, Queen's College, Oxford', *Domus*, November 1972, p. 516.

Rykwert, Joseph, 'The Constitution of Bohemia', *RES*, 1997, vol. 31, pp. 109–27.

Rykwert, Joseph, *The Judicious Eye, Architecture Against the Other Arts*, Chicago, IL: University of Chicago Press, 2008.

Saarinen, Aline B., *Eero Saarinen on His Work*, New Haven, CT: Yale University Press, 1962.

St John Wilson, Colin, and Williams, Edward, 'Sigurd Lewerentz', *The Architects Journal*, 1983, vol. 187, pp. 32–52.

Sadler, Simon, *Archigram: Architecture without Architecture*, Cambridge, MA: MIT Press, 2005.

Samuel, Flora, 'Suburban Self Build', *Field*, 2008, vol. 2, pp. 111–25

Saunders, William (ed.), *Judging Architectural Value: A Harvard Design Magazine Reader*, Minneapolis, MN: University of Minnesota Press, 2007.

Serlio, Sebastiano, *Sebastiano Serlio on Architecture, Vol. 1, Books I–V of 'Tutte l'Opere d'Architettura et Prospetiva'*, trans. Vaughan Hart and Peter Hicks, New Haven, CT: Yale University Press, 1996.

Schön, Donald M., *The Reflective Practitioner: How Professionals Think in Action*, Aldershot: Arena, 1995.

Schwarzer, Mitchell, 'The Emergence of Architectural Space: August Schmarzow's Theory of Raumgestaltung (Spatial Forming)', *Assemblage*, August 1991, p. 15.

Shaffer, E.S. (ed.), *Comparative Criticism: Walter Pater and the Culture of the fin-de-siècle*, Cambridge: Cambridge University Press, 1995.

Singley, Paulette, 'Roma Macchiata: The Stain of White', *Log*, Summer–Fall, 2007, vol. 10, pp. 129–36.

Snow, C.P. (Charles Percy), *The Two Cultures and the Scientific Revolution*, Cambridge: Cambridge University Press, 1959.

Stamp, Gavin, '"Stirling's Worth": The History Faculty Building', *Cambridge Review*, 30 January 1976, pp. 77–82.

Stamp, Gavin, 'The Durability of Reputation: The Cambridge History Faculty', *Harvard Design Magazine*, Fall 1997, p. 57.

Steadman, Philip, *Evolution of Designs: Biological Analogy in Architecture and the Applied Arts*, Cambridge: Cambridge University Press, 1979.

Steadman, Philip, *Architecural Morphology: An Introduction to the Morphology of Building Plans*, London: Pion, 1983.

Stirling, James, *James Stirling: Buildings & Projects 1950–1974*, London: Thames & Hudson, 1996.

Stokes, Adrian, 'The Sculptor Agostino di Duccio' [Extracted from the third of four essays on the Tempio Maletestino [sic] at Rimini], in T.S. Eliot (ed.), *The Criterion 1922–1939, Vol. IX, October 1929–July 1930*, London: Faber & Faber, 1967.

Stokes, Adrian, *Sunrise in the West*, New York and London: Harper & Brothers, 1927.
Stokes, Adrian, *The Quattro Cento and Stones of Rimini*, University Park, PA: The Pennsylvania State University Press, 2002.
Stokes, Adrian, *With All the Views. The Collected Poems of Adrian Stokes*, Manchester: Carcanet New Press, 1981.
Strauven, Francis, *Aldo van Eyck: The Shape of Relativity*, Amsterdam: Architectura & Natura, 1998.
Swoboda, Karl, *Peter Parler, Der Baukünstler und Bildhauer*, Wien: Verlag Anton Schroll, 1940.
Szarkowski, John, *Atget*, New York: Calloway, 2000.
Szarkowski, John, *The Idea of Louis Sullivan*, London: Thames & Hudson, 2000.
Szarkowski, John, *The Photographer's Eye*, London: Thames & Hudson, 2007 (first published 1966).
Sullivan, Louis, *The Autobiography of an Idea*, New York: Dover Press, 1956.
Tange, Kenzo, *Katsura: Tradition and Creation in Japanese Architecture*, New Haven, CT: Yale University Press, 1960.
Till, Jeremy, *Architecture Depends*, Cambridge, MA: MIT Press, 2009.
Thompson, Darcy Wentworth, *On Growth and Form* (1917), reprint, New York: Cambridge University Press, 1961.
Thorne, Martha and Amery, Colin (eds), *The Pritzker Architecture Prize: The First Twenty Years*, New York: Harry N. Abrams, Inc., Publishers in association with The Art Institute of Chicago, 1999.
Treib, Marc, *Sanctuaries of Spanish New Mexico*, Berkeley, CA: University of California Press, 1993.
Treib, Marc, 'To End a Continent: The Courtyard of the Salk Institute', *Journal of the Society of Architectural Historians*, September 2006.
Trump, David H., *Malta: Prehistory and Temples*, Malta: Midsea Books, 2004.
Unrau, John, *Looking at Architecture with Ruskin*, London: Thames & Hudson, 1978.
Upitis, Alise, 'Personality, Consumption and Architectural Pedagogy in the UK, 1958–1968', *arq*, 2008, vol. 12, pp. 3–4.
van der Bergh, Wim, *Luis Barragán: The Eye Embodied*, Maastricht: Pale Pink Publishers, 2006.
van Doesburg, Theo, *Principles of Neo-Plastic Art* (1925), trans. Janet Seligman, New York: New York Graphic Society, 1966.
Vastert, Eric, 'Specification of Visual Requirements for Regularity of Joints', *Architectural Science Review*, September 1998, vol. 41, pp. 99–104.
Vidler, Anthony, 'Losing Face: Notes on the Modern Museum', *Assemblage*, June 1989, p. 9.
Viollet-le-Duc, Eugène-Emmanuel, 'Architecte', in *Dictionnaire raisonné de l'architecture francaise du XIe au XVIe siècle*, Paris: Gründ, 1854–67.
Vitruvius, Pollio Marcus, *Ten Books on Architecture*, New York: Dover; and London: Constable, 1960.
Webster, Helena, 'The Analytics of Power: Re-presenting the Design Jury', *Journal of Architectural Education*, 2006, vol. 60, p. 3.
Westbrook, Adele and Yarowsky, Anne (eds), *Design in America: The Cranbrook Vision, 1925–1950*, New York: Harry Abrams, 1983.
Whitehill, Walter Muir, 'The Making of an Architectural Masterpiece: The Boston Public Library', *American Art Journal*, Autumn 1970, vol. 2, no. 2, pp. 13–35.
Wigley, Mark, 'How Old Is Young?', in *Famous: Perspecta 37, The Yale Architectural Journal*, Cambridge, MA: MIT Press, 2005, pp. 64–77.
Wildman, Stephen, *'My Dearest Place in Italy': Ruskin in Verona*, Lancaster University: Ruskin Library, 2005, Exhibition catalogue.
Willard, Russell L., 'Signed Buildings', *Pencil Points*, September 1929, vol. 10, pp. 633–4.
Wittkopp, Gregory (ed.), *Saarinen House and Garden: A Total Work of Art*, New York: Harry Abrams, 1995.
Woods-Marsden, Joanna, *Renaissance Self-portraiture: The Visual Construction of Identity and the Social Status of the Artist*, New Haven, CT: Yale University Press, 1998.
Wotton, Henry, *The Elements of Architecture*, 1624.
Zanco, Federica, (ed.), *Luis Barragán: The Quiet Revolution*, Milan: Skira, 2001.
Zenner, Marie-Therese (ed.), *Villard's Legacy: Studies in Medieval Technology, Science and Art in Memory of Jean Gimpel*, Aldershot: Ashgate, 2004.

Index

Page numbers in *italics* refer to figures.